“Hours like bright sweets in a jar”: Time and Temporality in Literature and Culture

"Hours like bright sweets in a jar": Time and Temporality in Literature and Culture

Edited by

Alicja Bemben and Sonia Front

CAMBRIDGE SCHOLARS

PUBLISHING

"Hours like bright sweets in a jar":
Time and Temporality in Literature and Culture,
Edited by Alicja Bemben and Sonia Front
Reviewed by Jadwiga Uchman, University of Łódź, Poland

This book first published 2014

Cambridge Scholars Publishing

12 Back Chapman Street, Newcastle upon Tyne, NE6 2XX, UK

British Library Cataloguing in Publication Data
A catalogue record for this book is available from the British Library

ISBN (10): 1-4438-6523-0, ISBN (13): 978-1-4438-6523-4

TABLE OF CONTENTS

List of Images

THE EDITORS' PREFACE

In his *Time and the Novel* Adam A. Mendilow has noted that the previous century underlines the significance of the study of time and temporality to the point of "time-obsession of the twentieth century which reveals itself in every art."[1] It is perhaps because writers and artists feel that "Only by getting to grips with the time-problem can they understand the meaning of living and acquire a true perspective of reality; only through their solution of the time-problem can they solve the problem of their art."[2] While certainly modernity has been fixated on time, in fact people have always attempted to solve its mystery. The aspect of time as an existential problem is highlighted, but not limited to, in the present volume. The collected papers undertake to investigate time and temporality from a number of interdisciplinary perspectives: literary or film studies, postcolonial theory, physics, philosophy, psychology, urban studies, history and gender studies. The wide spectrum of scholarly approaches encompasses chapters dealing with the convergences of time and human psyche, time and the body, time and memory, time and trauma, time and change, time and cultural reproduction, time and language, time and the city, and time and identity.

The volume is divided into two sections: "Frameworks" and "Readings." "Frameworks" offer some broader contexts and theoretical positions from which to approach the notions of time and temporality. "Readings," on the other hand, concentrate on close readings of individual literary texts or films from a variety of critical standpoints.

The book is opened by **Tomasz Burzyński** (University of Silesia, Poland) who proposes some methodological reflections on temporal anatomies in cultural studies and social sciences. He juxtaposes theories promoting the conceptualization of time as the linear progressive notion of socio-cultural development with non-linear notions of "multiple modernities" and "reflexive modernization," in which the pattern of history arises from the contingent and multidirectional "intermeshed plurality of events." He investigates the consequences of employing such temporally-oriented analyses.

---

[1] Adam Abraham Mendilow, *Time and the Novel* (New York: P. Neville, 1952), 12.

[2] Mendilow, *Time and the Novel*, 16.

Sonia Front (University of Silesia, Poland) traces the shifts in the notions and conceptualizations of time in western culture and their literary counterparts, representations and parallels. Examining the influences on modernist and postmodernist cultural and literary concepts of time, she emphasizes the scientific revolution performed by the new physics, and explores the representations of its concepts connected with temporality in British physics fiction/quantum fiction.

Katarzyna Ancuta (Assumption University, Bangkok) in her urban study of the city of Bangkok discusses the multiple temporalities of the city. In contrast to the dominant descriptions of Bangkok's time as frozen, halted by the eternal traffic jam, Ancuta sheds some light on what she calls "the second city," spectral and ephemeral in its being, which is realized through the ritualistic performance of its inhabitants.

The following two articles, opening the section of "Readings," are concerned with postcolonial temporalities, trauma and memory. **Katarzyna Nowak-McNeice**'s (University of Wrocław, Poland) point of departure in her essay is the assumption that Americans locate their sense of unique identity in the temporal amputation, chopping off their past and turning to the future. On the basis of Helen Hunt Jackson's novel *Ramona* (1884) Nowak-McNeice shows that the denied past demands the process of mourning, otherwise it is going to return hauntingly. The exorcism of the past is also performed by the characters in the postcolonial speculative short story "Griots of the Galaxy" (2004) by Andrea Hairston, analyzed by **Agnieszka Podruczna** (University of Silesia, Poland). The story rewrites the past from the point of view of the marginalized subject, offering disruptive narratives and polyphonic disjointed memories as the site of the construction of identity, which defies the colonizing linear master narrative.

Sławomir Konkol's (University of Silesia, Poland) paper does not stray from the subject of trauma and the spectral past, which is the focus of Graham Swift's *Waterland* (1983). Applying Lacan's theories, Konkol investigates the temporal collapse at the moment of the traumatic event and the protagonist's ceaseless attempts to assimilate and process the event through storytelling that would enable liberation from the entrapment of trauma.

Alicja Bemben (University of Silesia, Poland) scrutinizes temporal ruptures in three novels: time-travelling in *Seven Days in New Crete* (1949) by Robert Graves, the reversed worldline of *The Man Who Lived Backward* (1950) by Malcolm Ross and the forward and backward movement in time during the journey to prelapsarian wholeness in Alejo Carpentier's *The Lost Steps* (1953). All the strategies serve, in Bemben's understanding, as an active confrontation with mechanical time which strips the characters' lives of meaning.

In the following chapter, **Jacek Mydla** (University of Silesia, Poland) draws our attention to the irregular temporalities that precede modernity. He offers an exploration of human time in William Shakespeare's plays, concentrating on the intersection of time and language in *Richard II* (1595). Human concerns with time are not only voiced in language but also related to it, therefore, Mydla argues, if language is negotiable, then time is as well.

The concluding chapter by **Karen Heald** (School of Art, Design, Technology, Leeds Metropolitan University, UK) provides a short overview of complex female temporalities in film, emphasizing ephemeral and liminal psychological states, such as dreaming, imagination, states between wakefulness and sleep, reverie and the unconscious, and referring them to Julia Kristeva's notions of the semiotic *chora* and "women's time." In her trans-disciplinary project, Heald, a scholar and an artist herself, collaborated with creative practitioners, sociologists and scientists of different professions to finally encapsulate the results of the research in her poetic, painterly, antilinear "dream films."

What comes to the foreground in all the articles is the fundamental antilinearity and irregularity of the imaginary reconfigurations of time. If linearity resides in mechanical (clock) time, mathematical time, Christian time, some notions of progress, colonial master narrative as well as the human psychological arrow of time in its irreversible being-towards-death, then fictive temporalities manifest themselves as the continuous resistance against these linearities. Sense-making strategies always encompass disobedience against the inhuman rigidity of the mechanical time of institutions, work, government, power and the Hegelian realm of devouring Chronos, endeavouring to colonize the human psyche. Trying to resist this hegemony of linear time, literary, cinematographic and cultural practice enacts exploding temporalities to reflect the multifacetedness of psychological time, female time, trauma, temporal layers of the city, digital time, the time of the new physics, and the colonial experience from the point of view of the colonized. These types of experience always clash with linear time as though exposing its dehumanizing effect. In their multi-directionality of the sense of time, they can be visualized as "bright sweets in a jar."[3] The metaphor, borrowed for our title from Penelope Lively's novel *Moon Tiger* (1987), points to the simultaneity of the experienced moments of time. In our collection, various other metaphors and

---

[3] Penelope Lively, *Moon Tiger* (London: Penguin Books, 1988), 108.

figurations have arisen, which can be interpreted as the counterparts of Lively's "bright sweets in a jar": "multiple modernities" or "reflexive modernization," "polyhistory," "polyphony of memories," "time broke[n] in a disordered string," "polychronic time orientation" in the city and prelapsarian wholeness. Likewise, the articles in the two sections of our book can be treated like "bright sweets in a jar" themselves, by which we suggest that they be read without the imposed order (although, of course, the linearity of the form of a book forces an order).

From the problematizing of chronological time presented in this collection various notions and conceptualizations of rhizomatic temporalities emerge: "liquid time," "unbounded time," frozen time, "perpetual present"/"extended present," empty time (as opposed to quality time), chronotope, "schismatic temporality," "instantaneous time," deep time, fugal time, kaleidoscopic time, forking paths, a backward worldline, and others. All of them serve to underline the plurality, heterogeneity and instability of the experience of time as well as its constant reluctance to be subjugated in simple and unequivocal definitions. They also corroborate the constant interest of many thinkers in describing various dimensions of time experience and its philosophical implications.

As demonstrated, the papers in the present collection constitute a truly inter/trans-disciplinary approach to time and temporality, sometimes revealing unexpected intersections and results. In so doing, the volume fills a gap in criticism and research on temporality, as the previously published studies can usually be located within the boundaries of one discipline. We hope, therefore, that the volume will become the focus of critical attention of scholars, critics and students of the subject who look for a multiplicity of perspectives.

FRAMEWORKS

CHAPTER ONE

TIME AND CULTURAL PRACTICE: SOME METHODOLOGICAL REMARKS ON TEMPORALLY-ORIENTED ANALYSES IN CULTURAL STUDIES

TOMASZ BURZYŃSKI

> "Now, it is very remarkable that this is so extensively overlooked," continued the Time Traveler, with a slight accession of cheerfulness. "Really this is what is meant by the Fourth Dimension, though some people who talk about the Fourth Dimension do not know they mean it. It is only another way of looking at Time. *There is no difference between time and any of the three dimensions of space except that our consciousness moves along it.* But some foolish people have got hold of the wrong side of that idea."[1]

Introduction

It seems relatively easy to forget that time is an essential element of cultural practice and, by the same token, an indispensable component of culturally-oriented methodologies. The significance of temporal variables in cultural studies is very often lost in the maze of counterfactual theoretical options and academic disputes over the ontology of socio-cultural realities. Likewise, the methodological importance of time-centered analyses is very frequently forced to give way to more conventional investigations aiming to uncover the essence of cultural practice by means of referring to the binary opposition between language system and interpretation, as is the case with a plethora of structural-

[1] Herbert George Welles, *The Time Machine* (New York: Henry Holt and Company, 1898), 8.

functional and historico-interpretative approaches. It also means, to put it slightly otherwise, that the traditional methodologies of cultural studies are mostly absorbed with the idea and ontological status of human subjectivity (agency). The concept of time, therefore, is regarded as a secondary variable with reference to the notion of the Subject, *l'enfant terrible* of contemporary, post-structural cultural theories. Hence, the very idea of time is downplayed and veiled by frantic attempts to deal with the unnerving presence of the Subject, the notion which is despised as an illusory construction produced by systemic or structural constraints implicit in linguistic and social structures, or demonized as the Cartesian construct of a skillful interpreter and active appropriator of the exterior.

Concomitantly, the awareness of restrictions implicit in the aforementioned paradigms has motivated numerous scholars to construct theories aiming to overcome the conceptual tensions and contradictions implicit in the binary opposition between agency and structure. From this perspective, the displacement of temporal variables from cultural studies is seen as a result of being indifferent to a possibility of synthetizing the discourses of agency and structure in a coherent form of sequential and time-related order of cultural morphogenesis. Anthony Giddens' "theory of structuration" constitutes a forefront voice in the *heteroglossia* of contemporary synthetic discourses:

> The theory of structuration was worked out as an attempt to transcend, without discarding altogether, three prominent traditions of thought in social theory and philosophy: hermeneutics or 'interpretative sociologies', functionalism and structuralism. Each of these traditions, in my view, incorporates distinctive and valuable contributions to social analysis – while each has tended to suffer from a number of defined limitations.[2]

This article aims to discuss the role of temporally-oriented analyses in the context of methodologies typical of social sciences and cultural studies. The text traces methodological consequences of deploying temporally-oriented analyses in cultural studies and theories of culture and, more specifically, observes the theoretical ramifications of referring to time-related considerations and variables, as they are evident as the extensive proliferation of discourses based upon the notions of agency, cultural reproduction and morphogenesis.

---

[2] Anthony Giddens, *A Contemporary Critique of Historical Materialism* (London: McMillan, 1981), 26.

Time and Change

Being a meaningful component of social actions, time is also an indispensable element of cultural practice. This is not only to say that culture belongs to the order of reconstruction and remembering, and all cultural practices are processes *in statu nascendi* which utilize pre-existent symbolic, normative and axiological resources to re-deploy them in order to deal with a parade of counterfactual scenarios of the future. The temporality of culture is also anchored in the activist, agency-centred perspective on culture in which signs, symbols, norms and values are discussed as consequences and coefficients to individual and collective actions. In this particular view, culture, to put it otherwise, becomes automatically associated with the de Saussurean notion of the *parole* conceived as the on-going, action-related deployment of linguistic regulations. Yet, unlike Ferdinand de Saussure's theory, this overly activist perspective gives a methodological primacy to the agential sphere of language use and cultural practice which are both regarded as observable manifestations of societal *praxis*:

> In its different ways, it conceptualizes culture as interwoven with all social practices; and those practices, in turn, as a common form of human activity: sensuous human praxis, the activity thorough which men and women make history. ... The experiential pull in this paradigm, and the emphasis on the *creative and on historical agency*, constitutes the two elements in the humanism of the position outlined.[3]

The conventional schools of social theory and cultural studies have tended to dismiss this ontological, action-related nature of temporality by conceptualizing time as an objective variable, a form of "environment" in which all social actions and signifying practices take place.[4] This form of a derogative conceptualization has been promoted by the sociological paradigm of modernization and the legacy of evolutionist methodologies which teach us that social and cultural realities could be conceptualised in terms of a series of supposedly timeless snapshots which constitute the

---

[3] Stuart Hall, "Cultural Studies: two paradigms," in *Media, Culture and Society. A Critical Reader*, eds. Richard E. Collins, James Curran, Nicholas Garnham, Paddy Scannell, Philip Schlesinger, Colin Sparks (London: Sage Publications, 1986), 39, emphasis added.

[4] Anthony Giddens, *Central Problems in Social Theory. Action, Structure and Contradiction in Social Analysis* (Berkeley and Los Angeles: University of California Press, 1979), 202.

diachronic configuration of social development.[5] In this particular view, time is automatically associated with the notion of systemic change that transforms whole social and cultural organizations according to the pattern of linear, evolutionary progress conceived as a differentiation of social and cultural forms. As a consequence, the inherently changeable nature of society and culture becomes somehow forgotten and downplayed as something taken-for-granted. One, consequently, tends to forget that "ontologically, society is nothing else but change, movement and transformation, action and interaction, construction and reconstruction, constant becoming rather than stable being."[6]

Modernization theories seem to convey a relatively simplified view on temporality in which the concept of time becomes subsumed within the idea of linear, developmental social and cultural changes. In this sense, time – as the classical modernization framework postulates – is associated with the gradual transformations affecting subsequent states of socio-cultural systems, the alterations which have been quite effectively conceptualized as a sequential logic of change implying the existence of the traditional (pre-modern), early modern (industrial), late modern (post-industrial), and networked (information-reliant) forms of societal cohabitation and cultural integration.[7]

In this specific sense, time is endowed with an overly and overtly ideological significance as a variable which follows a predictable, developmental path heading towards the Hegelian end of history. Yet, this kind of reasoning may be seen as a gross simplification since the very idea of linear, progressive development is liable to be debunked:

> … history does not possess any metaphysical, substantive reality. Thus, in this image, the pattern of history is not superimposed or pre-established, but rather emerges out of [an] intermeshed plurality of events. Such a pattern is not treated as unique or singular, but rather emerges as the combined product of multiple sequences, overlapping and parallel, convergent and divergent, contradicting and complementing each other. It is not seen as a uniform and unidirectional process, but may change

---

[5] Giddens, *Central Problems*, 198.

[6] Piotr Sztompka, "The Trauma of Social Change. A Case of Postcommunist Societies," in *Cultural Trauma and Collective Identity*, eds. Jeffrey C. Alexander, Ron Eyerman, Bernard Giesen, Neil J. Smelser, Piotr Sztompka (Berkeley and Los Angeles: University of California Press, 2004), 155.

[7] Kazimierz Krzysztofek and Marek Stanisław Szczepański, *Zrozumieć rozwój. Od społeczeństw tradycyjnych do informacyjnych* (Katowice: Wydawnictwo UŚ, 2002), 36.

> direction, course and speed. It is not viewed as approaching any fixed, final goal, but is open-ended and contingent, allowing alternative scenarios.[8]

The notion of non-linear development seems to problematize the very concept of temporality. Non-linearity paves the way for a critical understanding that changes are not progressive in the sense of "moving ahead" according to the pre-established sets of social, economic or political coordinates. In this way, the very paradigm of modernisation is liable to be discredited since the developmental pattern assuming a fairly logical sequence of differentiation becomes dispersed into a parade of counterfactual modernization scenarios. This is, for instance, the case with the aforementioned idea of societal development conceived as a perfection of socio-cultural forms implying the gradual transformation of traditional agrarian communities into industrialised societies which, in turn, are doomed to be converted into post-modern (late modern) forms of networked society. This pattern has been problematized with the introduction of methodologies utilizing the theoretical constructs of "multiple modernities" or "reflexive modernization" in order to fulfil key explanatory functions. To listen to Ulrich Beck:

> This type of confrontation of the *bases of modernization* with the *consequences of modernization* should be clearly distinguished from the increase of knowledge and scientization in the sense of self-reflection and modernization. … Then "reflexive modernization" means the self-confrontation with the effects of *risk society* that cannot be dealt with and assimilated in the system of industrial society – as measured by the latter's institutionalised standards.[9]

The viewpoint, to put it even more methodologically, is tantamount to analyses of changes occurring *within* a given cultural system, not transformations *of* the whole system in question. Therefore, the former perspective is concerned, first and foremost, with the inherent dynamism of social and cultural institutions, with their innate potential for self-transformation. Consequently, cultural systems are perceived as entities in the state of fluidity, which allows for the gradual demystification of hidden developmental possibilities implicit in them.

---

[8] Piotr Sztompka, *Society in Action. The Theory of Social Becoming* (Cambridge: Polity Press, 1991), 71.

[9] Ulrich Beck, "The Reinvention of Politics: Toward a Theory of Reflexive Modernization," in *Reflexive Modernization. Politics, Tradition and Aesthetics in the Modern Social Order*, eds. Ulrich Beck, Anthony Giddens, Scott Lash (Stanford: Stanford University Press, 1994), 6, emphasis added.

Culture in the Process of Reproduction

When approached from the perspective of investigating changes occurring *within* socio-cultural systems, any form of temporally-oriented methodology is liable to face the problem of cultural reproduction indicating that culture is, at the end of the day, an amalgam of transient individual and collective actions as well as structural and institutional underpinnings of systemic perseverance.[10] Therefore, the notion of cultural reproduction is central to the very concept of culture because it implies its processual, active-reactive character which is embodied in mutually oriented signification processes and other types of reciprocal social actions:

> Culture is the production, reproduction, and transmission of relatively stable informational processes and their public representations, which are variously distributed in groups or social networks. The information is declarative and procedural, pertaining to ideas, beliefs, values, skills, and routinized practices as well as information about transmission process. The transmission occurs both between and within generations; moreover, the processes are shred unevenly, may be spread across non-localized groups, and may not be integrated.[11]

This inherent dynamism of cultural reality is experienced as the dialectic of continuity and change occurring within the systemic character of norms, values and signs of culture. Likewise, it is also discernible as a historical process manifesting itself as the *longue dureé* of institutional time as well as generalized changes of the whole system of culture. In this sense, as Chris Jenks declares, the very notion of culture "emerges from the noun "process," in the sense of nurture, growth and bringing into being."[12] Needless to say, this processual, inherently dynamic understanding of culture is etymologically intelligible and refers to the Latin expression *cultura animi* which evokes an idealistic, one may even say an Arnoldian,

---

[10] Orlando Patterson, "The Mechanisms of Cultural Reproduction," in *Handbook of Cultural Sociology*, eds. John R. Hall, Laura Grindstaff and Ming-Cheng Lo (London and New York: Routledge, 2010), 139-140.

[11] Patterson, "The Mechanisms of Cultural Reproduction," 139.

[12] Chris Jenks, "Introduction: The Analytic Bases of Cultural Reproduction Theory," in *Cultural Reproduction*, ed. Chris Jenks (London and New York: Routledge, 1993), 3.

conceptualization of culture as an activity of mental self-perfection, a process in which intellectual capacities are cultivated and tended.[13]

Moreover, the theory of cultural reproduction postulates that culture does not belong solely to the realm of recollection in which the heritage of past experiences establishes its own conceptual hegemony over the actuality of the present day. Important as it may be, cultural tradition constitutes a wavering construct, a mere simulacrum which is constantly reproduced and altered by acts of interpretation and processes of symbolic communication. In this way, the very process of cultural reproduction seems to be based upon the faculty of memory conceived as an agential alternation of the past. Let us quote the following passage *in extenso*:

> In addition to this reinterpretation *in toto* there must be particular reinterpretations of past events and persons with past significance. The alternating individual would, of course, be best off if he could completely forget some of these. But to forget completely is notoriously difficult. What is necessary, then, is a radical reinterpretation of the meaning of these past events or persons in one's biography. Since it is relatively easier to invent things that never happened than to forget those that actually did, the individual may fabricate and insert events wherever they are needed to harmonize the remembered with the reinterpreted past. Since it is the new reality rather than the old that now appears dominatingly plausible to him, he may be perfectly 'sincere' in such a procedure – subjectively, he is not telling lies about the past but bringing it in line with the truth that, necessarily, embraces both present and past.[14]

The alternation of the past supplies a sense of temporal continuity which is indispensable for the process of cultural reproduction. This critical interplay between continuity and change is, for instance, a central element of Pierre Bourdieu's theory of the *habitus* which indicates that all actions aiming to produce elements of symbolic culture are, at the same time, aimed to re-produce it in a sense of being anchored in the historically established modes of taste, and normative evaluation or moral judgement.[15] In this particular sense, the *habitus* functions as "systems of durable, transposable dispositions, structured structures predisposed to

---

[13] Interestingly enough, a verb form "to culture" is very often forgotten and replaced by its contemporary counterparts.

[14] Peter Ludwig Berger and Thomas Luckmann, *The Social Construction of Reality. A Treatise in the Sociology of Knowledge* (London: Penguin Books, 1991), 180.

[15] Pierre Bourdieu, *The Logic of Practice* (Cambridge: Polity Press, 1992), 53.

function as structuring structures"[16]; that is, the *habitus* operates as a systemic underpinning which supplies symbolic actions with their discursive, normative and axiological orientations.

The process-oriented understanding of culture paves the way for an overtly activist methodology in which cultural reproduction is perceived in terms of societal processes of conflict and negotiation taking place with regard to diverse components of symbolic culture. Being oriented at the persistency of such cultural constituents as norms, values or symbols, the process of cultural reproduction resembles, as a matter of fact, a discursive struggle taking place over the dominant or, as Stuart Hall puts it, preferred meanings.[17] Socio-cultural systems or institutions, consequently, resemble constantly changing matrixes or networks of inter-subjectively granted phenomena which are being constituted as a result of myriads of intelligible processes of communication as well as interaction undertaken by knowledgeable agents.

The Temporality of Cultural Reproduction Processes

When converted into the conceptual reality of cultural studies, the ontological aspect of temporality suggests that the reproduction of culture involves three closely intertwined orders of temporality.[18] The first aspect of temporality is the reversible *durée* of daily routines. This is a temporal order which is constituted by the compulsion of constant repetition. Repetitiveness is, in this case, conceived as a temporal foundation of cultural order and, in the long run, an anchorage of relatively stable trust expectations vested in the reliability of social, economic or political systems. This is, to put it otherwise, the cornerstone of "ontological security"[19] indicating that trust vested in the continuity of things may function in terms of an "emotional inoculation against existential anxieties."[20] This, consequently, gives rise to a natural expectation that uncertainties and contingencies may be, to use a phenomenological nomenclature, bracketed off.

---

[16] Bourdieu, *The Logic of Practice*, 53.

[17] Stuart Hall, "Encoding, Decoding," in *The Cultural Studies Reader*, ed. Simon During (London and New York: Routledge, 1994), 98.

[18] Anthony Giddens, *The Constitution of Society. Outline of the Theory of Structuration* (Berkeley and Los Angeles: University of California Press, 1984), 34-36.

[19] Anthony Giddens, *Modernity and Self-Identity. Self and Society in the Late Modern* Age (Stanford: Stanford University Press, 1991), 36-42.

[20] Giddens, *Modernity and Self-Identity,* 39.

The reversible time of daily routines constitutes a form of temporality which is deprived of its own *telos*. "Daily life has a duration, a flow but it does not lead anywhere … time is constituted only in repetition."[21] The reversible *durée* of daily routines is reflected *mutatis mutandis* by external cultural realties. One quite naturally relies on the continuous flow of social order gesturing to the critical expectance that cultural resources are also endowed with an inherently repetitive character.

A strikingly dissimilar order of temporality is discussed when the problem of human ontogenesis becomes evoked. The overall course of human life institutes the irreversible time of human ontogenetic existence. As opposed to the reversible time of daily routines, which corresponds to the ontological security attained at the cost of compulsive repetition, the irreversible time of being-towards-death constitutes the symbolic foundation of individual identity. Psychologically speaking, this particular order of temporality is reflected by the typically human need for being meaningful; namely, for creating a personal self-narrative that would explain the supposedly meaningless and repetitive flow of daily experience by evoking a vision, no matter how fragile it may be, of personal self-fulfilment. Likewise, this form of temporality evokes an entirely different type of trust expectations. In this context, agents are obliged to vest their trust expectations in the life-long continuity of social and cultural institutions. In other words, this aspect of ontological security is related to a belief that one's existence will be safeguarded by the temporal coherence of successive states of society and culture.

The mutually exclusive expectations of repetitiveness and purposefulness are yoked together in a form of the reversible institutional time. This specific axis of temporality connotes as an essential element of trans-personal institutions of culture, a set of persistent structural rules and resources rendering possible the repetitiveness of daily routines and the purposefulness of human existence. The institutional time gestures to the historical rhythm of social and cultural institutions which indicates the supra-individual patterns of persistency and change evident, for instance, in the case of modernization processes.

This triadic typology is, in fact, useful only for purely analytic purposes. The very concept of cultural reproduction is based upon the critical premise that the process assumes the unity of the aforementioned orders of temporality:

> The reversible time of institutions is both the condition and the outcome of the practices organised in the continuity of daily life, the main substantive

[21] Giddens, *The Constitution of Society*, 35.

> form of the duality of structure. It would not be true, however, as I have already mentioned, to say that the routines of daily life are [the] 'foundation' upon which the institutional forms of societal organisation are built in time-space. Rather, each enters into the constitution of the other, as they both do into the constitution of the acting self. All social systems, no matter how grand or far-flung, both express and are expressed in the routines of daily life, mediating the physical and sensory properties of the human body.[22]

Agents rely upon the existence of the repeatable *milieu* constituting the day-to-day practice, as well as their "grand performance" representing the irreversible flow of activities heading towards the end of earthly matters. The temporal reality of the daily routine as well as the irreversible time of being-towards-death relate to each other and, consequently, create – to use Fernand Braudel's terminology[23] – the historically defined *longue durée* of social and cultural institutions. This reversible time of trans-personal phenomena constitutes the sphere of constraints as well as facilitations fostering and enabling the course of daily routine by supplying it with the rules and resources of systemic reproduction. By the same token, institutional time renders possible the sense of ontological security experienced with reference to the irreversible time of being-towards-death.

The triadic model of temporality is a useful analytic tool for observing socio-cultural processes and phenomena that are bound by time-related practices, as is the case of tradition. The cultural significance of tradition – to put it in the most general context – alludes towards a distinct strategy of organising and controlling time.[24] This concept ought to be understood in terms of a trans-individual phenomenon and, since individual traditions are not conceivable, it is tantamount to the *longue durée* of social and cultural institutions. Hence, traditions may be subsumed within the idea of collective memory or collective consciousness. The supra-individual aspect of the phenomenon gestures towards an idea that all traditions involve some degree of multilateral organisation of interpersonal relations as well as reciprocity. In this specific way, the term reflects the existence of social consent with reference to the accepted hierarchies of norms and values.

---

[22] Giddens, *The Constitution of Society*, 36.

[23] Fernand Braudel, *On History* (Chicago: University of Chicago Press, 1980).

[24] Anthony Giddens, "Living in a Post-Traditional Society," in *Reflexive Modernisation. Politics, Tradition and Aesthetics in the Modern Social Order*, eds. Ulrich Beck, Anthony Giddens and Scott Lash (Stanford: Stanford University Press, 1994), 62.

> Tradition, it might be said, is an orientation to the past, such that the past has a heavy influence or, more accurately put, is made to have a heavy influence over the present. Yet, clearly, in a certain sense at any rate, tradition is also about the future, since established practices are used as a way of organising future time.[25]

In the context of the aforementioned triadic order of temporality, the reign of tradition becomes a collective medium of memory rendering possible the organisation of reversible time associated with day-to-day activates. As far as pre-modern cultures are concerned, the authority of the past goes hand in hand with the organisation of activities concerning the totality of day-to-day pursuits, especially when those tend to involve coping with uncertainties and risks.[26] Moreover, the media of collective memory, such as myths, folktales, or legends, supply actors with solutions to problems of eschatology: they help re-organise human life so that it contains more sense and order in its process of oscillating for eternity. Running parallel to these statements is the appreciation of tradition as a living phenomenon; that is, a cultural institution that has no other conceivable mode of existence, than through the actions undertaken both in the scope of daily routines and human life in its totality.

Agency and Structure

The temporal qualities of cultural reproduction processes pave the way for a methodological dilemma which is very frequently presented as a duality of agency and structure. The concepts, to cut a long story short, are subsumed within a discourse that aims to understand social and cultural realities from the perspective of an interplay between persistent and societally superintended rules of systemic reproduction (the aspect of "structure") and, on the other hand, creative drives which constitute momentary aberrations to the superintended rules of persistence and introduce changes to the otherwise systematic flow of socio-cultural phenomena (the aspect of "agency").[27]

---

[25] Giddens, "Living in a Post-Traditional Society," 62.

[26] Magic and myths seem to constitute a pre-modern response to, to use contemporary discourses, problems of occupational safety and risk management. See Bronisław Malinowski, "Myth in Primitive Psychology," in *The Myth and Ritual Theory. An Anthology*, ed. Robert A. Segal (Oxford: Blackwell Publishers, 1998), 173-179.

[27] Tomasz Burzyński, "The Surplus of Structure. Towards the Morphogenetic Approach to Cultural Studies," in *The Surplus of Culture. Sense, Common-sense,*

When translated into the methodology of cultural studies, the duality of agency and structure gives rise to an inherently dialectic view on cultural realities which are conceptualized as compound totalities fusing forces of structural determinism with the typically agential propensity to introduce innovations to the already existing schemata of cultural resources:

> On the one hand culture provides a pool of resources of action that draws form it the values to set its goals, the norms to specify the means, the symbols to furnish it with meaning, the codes to express its cognitive content, the frames to order its components, the rituals to provide it with continuity and sequence and so forth. In brief, culture supplies action with axiological, normative and cognitive orientation. In this way it becomes a strong determining force, releasing, facilitating, enabling, or, as the case might be, arresting, constraining, or preventing action. On the other hand, action is at the same time creatively shaping and reshaping culture, which is not God-given, constant, but rather must be seen as an accumulated product, or preserved sediment of earlier individual and collective action.[28]

The duality of culture represents the innate potentiality of self-change or self-elaboration by means of actions undertaken by knowledgeable agents. In this specific context, the term "morphogenesis" is very often deployed to refer "to those processes which tend to elaborate or change a system's given form, state or structure."[29] From a purely morphogenetic perspective, culture is regarded as a changeable entity whose dynamism is rooted in actors' agency, the typically human *élan vital* that aims to reconstruct surrounding social and cultural environments. Being embedded in the theory of action, the theory of morphogenesis is not deprived of attempts to investigate into the external, structural and systemic conditionings of human agency. In this case, the reproduction of culture is perceived as both enabled and conditioned by individuals' interpretative actions, as well as by the forces of structural determination constituting the axiological, normative, symbolic, or discursive bases for the activities in question.

Despite its considerable and multifaceted methodological impact, the very idea of the structural-agential duality is a relatively new one in the humanities and social sciences. By contrast, the classical, deeply rooted in

---

Non-sense, eds. Ewa Borkowska and Tomasz Burzyński (Newcastle upon Tyne: Cambridge Scholars Publishing, 2011), 223-234.

[28] Piotr Sztompka, *Trust. A Sociological Theory* (Cambridge: Cambridge University Press, 1999), 3-4.

[29] Margaret S. Archer, *Realist Social Theory: the Morphogenetic Approach* (Cambridge: Cambridge University Press, 1995), 166.

the legacy of the nineteenth century philosophies, methodologies rest upon the formative notion suggesting that the dimensions of human agency and systemic and structural constraints should be treated in terms of a binary opposition. In this conventional sense, the idea of structures (or systems) is conceptualized as a depository of persistence, a relatively stable universe of objective relations which, as the case of Durkheimian/de Saussurean structural-functional approach teaches us, are constituted in the *longue durée* of socio-cultural evolution, not in the protean reality of everyday action or language use. On the other hand, the category of human activity is deposited in the hermeneutic and agential capacities of human consciousness whose intentional grasp functions to deploy subjective meanings to transitory acts of perception.

The contemporary insights into the nature of cultural practice[30] suggest that the two conventional perspectives, nevertheless, offer dilemmas as inexplicable as squaring the circle. Consequently, there is a need for developing a more synthetic discourse that would leave these debates behind:

> In the model of social becoming, the levels of structure in operation and of agents in actions will be treated neither as analytically separable nor as mutually reducible. Instead a third, intermediate level will be postulated, and it will be claimed that it represents the only true substance of social reality, the specific social fabric. If we think of any empirical event or phenomenon in a society, anything that is actually happening, is it not always, without exception, a fusion of structures and agents, of operation and action? Show me an agent who is not enmeshed in some structure. Show me a structure which exists apart from individuals. Show me an action which does not participate in societal operation. Show me societal operation not resolving into action. There are neither structureless agents nor agentless structures.[31]

The fusion of structures and agents is rendered possible by means of taking temporal variables into account. Time, as Margaret S. Archer observes, is an inevitable element of theory-making since "structure and action operate over different time periods … structure logically predates the actions that transform it and structural elaboration [morphogenesis –

[30] This perspective is most evident in Anthony Giddens' theory of structuration, Piotr Sztompka's theory of social becoming and Margaret S. Archer's morphogenetic approach.

[31] Sztompka, *Society in Action*, 91-92.

T.B.] logically post-dates those actions."[32] In this way, the introduction of time gives rise to a methodology which reflects the dynamic, inherently changeable nature of social and cultural realities and explains it in terms of the alternating sequences of structural determination and structural elaboration.

This temporally-oriented duality of agency and structure paves the way for an image of socio-cultural ontology that rests upon five interrelated premises: 1) structuralism, 2) creativism, 3) processualism, 4) possibilism and 5) reflexiveness.[33] These tenets constitute a theoretical basis for providing a more informed perspective on cultural phenomena in which time-related variables assume crucial explanatory functions with reference to observing changes taking place *within* the system of culture. In this sense, furthermore, the factor of time is indispensable in order to understand the nature of mutual relationships occurring between human action and interpretation, and structural rules and resources existing in the shape of constraints and facilitations implicit in systems of signs, norms, or values. Needless to say, the ultimate goal is to suggest an ontology that would be liberated both from the idiosyncrasies of interpretative theories as well as the structural-functional paradigm's scientific rigidity.

> It is a world in which reflexive individuals are seen as creatures and creators as the same time, social wholes as fluid relational networks humanly made, but also affecting people, and historical processes as the stream of incessant interplay of emergence and determination, in the course of which both individuals and society undergo cumulative transformations.[34]

Despite its deceptive name, the first premise rests upon a new interpretation of structuralism which postulates the necessity of perceiving socio-cultural realities in terms of patterns or regularities that may be observed within diverse empirical phenomena. In this context, cultural structures and systems are deprived of their nature *sui generis* and, consequently, are conceptualised in terms of societal aggregates comprising organised cultural practices. In this sense, culture is perceived as displaying certain structural proprieties; that is, regular patterns of symbolic-oriented actions which are undertaken in a relatively conscious manner. As a result, methodologies should be concerned, first and foremost, with observing

---

[32] Margaret S. Archer, *Taking Time to Link Structure and Agency* (New Delhi: Eleventh World Congress of Sociology, 1986), 22. Quoted after: Sztompka, *Society in Action*, 106.
[33] Sztompka, *Society in Action*, 51-86.
[34] Sztompka, *Society in Action*, 86.

temporal regularities in the socio-cultural reality without granting them any mode of operation which is pre-existent with reference to actions undertaken by agents. This line of explanation, consequently, follows from the dimension of human actions towards the realm of emergent socio-cultural structures.

The premise of "creativism" refers to the dimension of creative drives implicit in agents perceived as active and goal-oriented participants in socio-cultural environments. In this context, individuals are inherently endowed with an impulse to act, and with an innate energy to initiate actions. Yet, at the same time, they refer to the external reality as a sphere of incentives and disincentives facilitating or restraining undertaken actions. In other words, agents are, to a certain extent, constituted and constructed within social and linguistic structures. However, at the same time, these systems function as the main vehicles of intentional, reflective creativity typical of human beings. From this perspective, an individual resembles the *homo creator*, an entity that uses encountered circumstances or objects in order to produce innovations. Consequently, in the context of operating agents, the creation made *ex nihilo* constitutes a totally illusory idea.

The premise of "processualism" refers to the nature of changes which take place with regard to society and culture understood as complete systems. As a product of human *praxis*, history is conceived in terms of an amalgamation of cumulated social actions and, consequently, is perceived as being deprived of its final objective. Nevertheless, the process of historical changeability does not constitute a chaotic universe of phenomena that manifest themselves in a random or accidental manner. It is characterised by a number of supra-individual rules constituting its sequential logic, which could be represented as the reversible time of institutions.

> Thus, in this image, the pattern of history is not superimposed or pre-established, but rather emerges out of the intermeshed plurality of events. Such a pattern is not treated as unique or singular, but rather emerges as the combined product of multiple sequences … .[35]

In other words, in spite of hopes expressed by enlightened Western intellectuals, the pattern of modernization is not a historical necessity and underdeveloped countries face plenty of possible routes towards modernisation, not only the one that has been conceived *ex post*, on the basis of historical experiences of the West. Hence, the notions of "multiple

[35] Sztompka, *Society in Action*, 71.

modernities" and "reflexive modernisation"[36] have been coined in order to put an emphasis on the non-linear, non-teleological process of socio-cultural development.

The premise of "possibilism" is based upon an assumption that actions participating in the re-production of the common socio-cultural habitat are simultaneously determined by their external circumstances. In a way, the assumption is a logical ramification of the aforementioned theses. Individuals are creators of the external socio-cultural environment, their agential potentialities become accumulated as patterns of historical changeability, yet history provides structural circumstances for human actions.

Finally, the duality of structure and agency assumes the reflexivity of social actors; that is, the intentionality of action which is a core aspect of human agency:

> 'Reflexivity' hence should be understood not merely as 'self-consciousness' but as the monitored character of the ongoing flow of social life. To be a human being is to be a purposive agent who both has reasons for his or her activities and is able, if asked, to elaborate discursively about those reasons (including lying about them).[37]

The idea of reflexivity is, nonetheless, strikingly dissimilar with reference to the structural-functional perspective on human subjectivity. Hence, it is the ideology heralding the demise of the subject – no matter how provocative it may be – that situates individuals as being irrelevant to the general construction of society and culture: "intentions or actions of human subjects, whether individual or collective, can easily be disposed of as irrelevant to the structural properties of the system."[38]

Conclusions

The importance of time-related considerations and variables in the conceptual and methodological framework of cultural studies and social sciences may be perceived in terms of synthetic discourses that aim to bind the dimension of agency and structure in a form of cultural ontology that

---

[36] Ulrich Beck, "Preface," in *Reflexive Modernization. Politics, Tradition and Aesthetics in the Modern Social Order*, eds. Urlich Beck, Anthony Giddens and Scott Lash, (Stanford: Stanford University Press, 1994), p. 6.

[37] Giddens, *The Constitution of Society*, 3.

[38] Andrew Miller and Jeff Browitt, *Contemporary Cultural Theory* (London, New York: Routledge, 2002), 98.

responds to the problem of cultural practice from a perspective of cultural reproduction and morphogenesis. In this context, time functions as a methodological keystone, the "vanishing mediator," to use Frederic Jameson's terminology,[39] between the systemic rules of persistence and the agential moments of innovation and creativity, thereby introducing changes to the supposedly static socio-cultural structures. Hence, the introduction of temporal analysis to cultural studies results in a new methodology in which the idea of culture is effectively conceptualized as a flow of on-going cultural practices relating structural rules and resources to individual and collective actions within the framework of cultural reproduction.

References

Archer, M. S., 1986, *Taking Time to Link Structure and Agency*, New Delhi: Eleventh World Congress of Sociology.

—. 1995, *Realist Social Theory: the Morphogenetic Approach*, Cambridge: Cambridge University Press.

Beck, U., 1994, "The Reinvention of Politics: Toward a Theory of Reflexive Modernization," in Beck, U., Giddens, A., Lash, S., eds., *Reflexive Modernization. Politics, Tradition and Aesthetics in the Modern Social Order*, Stanford: Stanford University Press, 1-55.

—. 1994, "Preface," in Beck, U., Giddens, A., Lash, S., eds., *Reflexive Modernization. Politics, Tradition and Aesthetics in the Modern Social Order*, Stanford: Stanford University Press, vii-viii.

Berger, P. L., and Luckmann, T., 1991, *The Social Construction of Reality. A Treatise in the Sociology of Knowledge*, London: Penguin Books.

Bourdieu, P., 1992, *The Logic of Practice*, Cambridge: Polity Press.

Braudel, F., 1980, *On History*, Chicago: University of Chicago Press.

Burzyński, T., 2011, "The Surplus of Structure. Towards the Morphogenetic Approach to Cultural Studies," in Borkowska E., and Burzyński, T., eds., *The Surplus of Culture. Sense, Common-sense, Non-sense*, Newcastle upon Tyne: Cambridge Scholars Publishing, 223-234.

Giddens, A., 1979, *Central Problems in Social Theory. Action, Structure and Contradiction in Social Analysis*, Berkeley and Los Angeles: University of California Press.

[39] Frederic Jameson, *The Ideologies of Theory. Essays 1971-1986, vol. 2: Syntax of History* (Minneapolis: University of Minnesota Press, 1988), 3.

—. 1981, *A Contemporary Critique of Historical Materialism*, London: McMillan.
—. 1984, *The Constitution of Society. Outline of the Theory of Structuration*, Berkeley and Los Angeles: University of California Press.
—. 1991, *Modernity and Self-Identity. Self and Society in the Late Modern Age*, Stanford: Stanford University Press.
—. 1994, "Living in a Post-Traditional Society," in Beck, U., Giddens, A., Lash, S., eds., *Reflexive Modernization. Politics, Tradition and Aesthetics in the Modern Social Order*, Stanford: Stanford University Press, 56-109.
Hall, S., 1986, "Cultural Studies: two paradigms," in Collins, R., Curran, J., Garnham, N., Scannell, P., Schlesinger, P., and Sparks, C., eds., *Media, Culture and Society. A Critical Reader*, London: Sage Publications, 33-48.
—. 1994, "Encoding, Decoding," in During, S., ed., *The Cultural Studies Reader*, London and New York: Routledge, page number.
Jameson, F., 1988, *The Ideologies of Theory. Essays 1971-1986, vol. 2: Syntax of History*, Minneapolis: University of Minnesota Press.
Jenks, Ch., 1993, "Introduction: The Analytic Bases of Cultural Reproduction Theory," in Jenks, Ch., ed., *Cultural Reproduction*, London and New York: Routledge, 1-16.
Krzysztofek, K., and Szczepański, M. S., 2002, *Zrozumieć rozwój. Od społeczeństw tradycyjnych do informacyjnych*, Katowice: Wydawnictwo UŚ.
Malinowski, B., 1998, "Myth in Primitive Psychology," in Segal, R. A., ed., *The Myth and Ritual Theory. An Anthology*, Oxford: Blackwell Publishers, 172-179.
Miller, A., and Browitt, J., 2002, *Contemporary Cultural Theory*, London, New York: Routledge.
Patterson, O., 2010, "The Mechanisms of Cultural Reproduction," in Hall, J. R., Grindstaff, L., and Lo, Ming-Cheng., eds., *Handbook of Cultural Sociology*, London and New York: Routledge, 139-151.
Sztompka, P., 1991, *Society in Action. The Theory of Social Becoming*, Cambridge: Polity Press.
—. 1991, *Trust. A Sociological Theory*, Cambridge: Cambridge University Press.
—. 2004, "The Trauma of Social Change. A Case of Postcommunist Societies," in Alexander, J. C., Eyerman, R., Giesen, B., Smelser, N. J., Sztompka, P., eds., *Cultural Trauma and Collective Identity*, Berkeley and Los Angeles: University of California Press, 155-195.

Welles, H. G., 1898, *The Time Machine*, New York: Henry Holt and Company.

CHAPTER TWO

TEMPORALITY IN BRITISH QUANTUM FICTION: AN OVERVIEW

SONIA FRONT

The notions and conceptualizations of time in western culture have undergone many shifts since the nineteenth century – they have been influenced by the Industrial Revolution, the scientific revolution connected with the new physics, the rise of a worldwide capitalist economy, and the media. In realist literature time figured as the background to the presented events to be problematized and achieve theoretical and thematic centrality only in the twentieth century. A chronological presentation of events in a realist novel gives way to disruptive temporalities in modernism although there are some texts of that type that precede modernity. The examples can be Geoffrey Chaucer's *The Canterbury Tales* with two parallel temporalities: that of "calibrated chronometrical time" and that of nature, or Lawrence Sterne's *The Life and Opinions of Tristram Shandy* (1759) with two temporal levels: of the story and of narrative discourse, with the discourse disrupting the story almost beyond the point of reconstructability.[1] This chapter constitutes an overview of the modern conceptualizations of time and the specific literary strategies to address them focusing on the literary responses to the new physics.

The Second Industrial Revolution in the latter half of the nineteenth century, with the introduction of the assembly line and steam power, the inventions of the radio, telephone and telegraph as well as the railway led to the standardization of time in the form of time zones set from Greenwich zero time.[2] The writers of the period often confronted the

---

[1] See Russell West-Pavlov, *Temporalities* (New York: Routledge, 2013), 13-14, 90-91.

[2] See Gerald James Whitrow, *Time in History. Views of Time from Prehistory to the Present Day* (Oxford, N.Y.: Oxford University Press, 1989), 152-169 and

regime of mechanical time and its tension with private time, e.g. Aldous Huxley in *Brave New World* (1932) or Virginia Woolf in *Mrs. Dalloway* (1925). While Huxley depicts Fordist time management, in Woolf's novel Big Ben as well as St Margaret and Clarissa's home clocks, whose striking is interspersed in the action, symbolize the masculinist time and the masculinist system of values, such as the will to power and authoritarianism.[3] Although Woolf presents the events of one day, she only seemingly constructs a chronological plot – in fact, she employs a back-and-forth technique: the plot "moves forward in the present day, it continually jumps back to earlier times, covering a segment of the past here, a segment of the past there, but never filling in all the gaps," and thus erases the boundary between the past and the present.[4] With the masculinist mechanical time of Big Ben, Woolf juxtaposes the feminine way of experiencing the world. One of the objectives of structuring narratives in an antilinear way is thus the feminist critique of the patriarchal authoritative version of history and the depiction of 'herstory' and gendered time.

Modernist and postmodernist notions of time were also influenced by the scientific revolution which depended on relativity theory and quantum mechanics. It resulted in the "paradigm shift"[5] from Newtonian physics to the Einsteinian relativistic worldview at macroscopic scales, and from classical mechanics to quantum mechanics at microscopic scales. It was accompanied by a shift in perspective also in mathematics, linguistics, philosophy, art, cinema and literature. The impact of the new physics upon metaphysics and aesthetics was unavoidable as it altered our notions of the universe, space and time. Victorian realism and rationalism were substituted by reality which became 'unrealistic,' that is non-commonsensical: relative, irregular, subjective, uncertain and fragmentary. Yet, it cannot be said that it was scientific discoveries alone that contributed to the "paradigm shift." The process is not causal – it is rather a result of a broader cultural *Zeitgeist*, a change of worldview taking place

---

Chapter 5 "Times," in John Urry, *Sociology Beyond Societies: Mobilities for the Twenty-First Century* (New York: Routledge, 2000), 105-130.

[3] Jo Alyson Parker, *Narrative Form and Chaos Theory in Sterne, Proust, Woolf and Faulkner* (NY: Palgrave Macmillan, 2007), 100. See also Paul Ricoeur, *Time and Narrative*, vol. 2, trans. Kathleen McLaughlin and David Pellauer (London, Chicago: The University of Chicago Press, 1985), 106-110.

[4] Parker, *Narrative Form and Chaos Theory in Sterne, Proust, Woolf and Faulkner*, 101.

[5] See Thomas Kuhn, *The Structure of Scientific Revolutions* (Chicago: University of Chicago Press, 1996).

in various fields simultaneously. N. Katherine Hayles describes this as a "climate, rather than direct borrowing or transmission, that is the underlying force guiding intellectual enquiry. ... the interactions are always mutual: the cultural matrix guides individual enquiry at the same time that the enquiry helps to form, or transform, the matrix."[6] It can, therefore, be stated that literary fiction has paralleled, foreshadowed or responded to the conceptual revolution, attempting to create metaphors and tropes to reflect a new notion of reality, including a new conception of time.

Relativity theory in fiction

Early modernist novels depict mainly psychological time and to do that, they often resort to relativity theory as their formal and/or thematic framework. Relativity is most often appropriated in novels as relativism. The literary interpretations of relativity take the form of novels employing a few narrative points of view to retell the same situation(s), which illustrates, in Wyndham Lewis's words, the "highly-complex *disintegrated* world of private 'times' and specific amputated 'spaces.'"[7] The omniscient narration is rejected as it is considered as so biased and unreliable as the characters. This distrust of the authorial omniscience contributed to a mass shift in the contemporary novel from the third person narration to the first person one. The literary interpretations of relativity theory include such modernist novels as, for example, Lawrence Durrell's *Alexandria Quartet* (1957-60), or James Joyce's *Ulysses* (1922) where in the part entitled "Wandering Rocks" the same set of events is depicted from a variety of perspectives.[8] Apart from that, Joyce satirizes relativity and points out its inability to depict human experience. What is significant in a novel of this kind is that none of the points of view can be treated as the dominating one; all of them are equally 'true.' Also, a temporal sequence emerges from the juxtaposition of simultaneous voices, and the different points of

---

[6] N. Katherine Hayles, *The Cosmic Web. Scientific Field Models and Literary Strategies in the Twentieth Century* (Ithaca and London: Cornell University Press, 1984), 22-23.

[7] Wyndham Lewis, *Time and Western Man* (Boston: Beacon Press, 1957), 417.

[8] Stephen Kern, however, attributes the narrative technique in *Ulysses* to the influence of cinema and its montage, and the objective of surmounting the chronological time of literature. See his *The Culture of Time and Space, 1880-1918: With a New Preface* (Cambridge, Massachusetts, London, England: Harvard University Press, 2003), 77, 149.

view complement each other and contribute to a fairly consistent story[9] which can be put together from the various pieces of the jigsaw puzzle.

Mikhail Bakhtin has borrowed Einstein's term "space-time" to construct his notion of a literary chronotope where he underlines the mutual interaction between time and space: "In the literary artistic chronotope, spatial and temporal indicators are fused into one carefully thought-out, concrete whole. Time, as it were, thickens, takes on flesh, becomes artistically visible; likewise, space becomes charged and responsive to the movements of time, plot and history. The intersection of axes and fusion of indicators characterizes the artistic chronotope."[10] While for Einstein space-time constitutes a single continuum, Bakhtin maintains that multiple chronotopes, that is space-time continuums, can clash in a novel. Both Einstein and Bakhtin recognize time as flexible, subjective and fused with space.

While Einstein himself did not suggest the connection between time and consciousness, sometimes it is mistakenly attributed to him. According to relativity theory, the duration of an event depends on the observer's velocity and reference frame, from which the measurements are taken. It is not necessary for the observer to have a mind. The literary explorations of the connection between time and consciousness have been influenced by philosophy – Henri Bergson's philosophy of duration, Sigmund Freud's psychoanalysis, complicating the relationship between inner and outer life, William James's psychology of time, Martin Heidegger's meditation on being-in-time and Edmund Husserl's phenomenology of internal time-consciousness. From the discrepancy between internal and external time ensues, in Derrida's interpretation, the madness of time. Derrida claims that time is "out of joint,"[11] off its hinges, bemused, entangled, mad, as it is haunted by its otherness or specter. Time is dismembered in its present and cannot achieve it as entirety.[12] The

---

[9] Ursula K. Heise, *Chronoschisms. Time, Narrative and Postmodernism* (Cambridge: Cambridge University Press, 1997), 50-51.

[10] Mikhail Mikhailovich Bakhtin, "Forms of Time and of the Chronotope in the Novel," in *The Dialogic Imagination: Four Essays by M. M. Bakhtin*, ed. Michael Holquist, trans. Caryl Emerson and Michael Holquist (Austin: University of Texas Press, 1981), 84.

[11] Derrida refers to *Hamlet* here: "The time is out of joint. Oh cursed spight, /That ever I was born to set it right!" William Shakespeare, *Hamlet* (Irvine, CA: Saddleback Educational Publ., 2003), 25.

[12] Jacques Derrida, "The Time is out of Joint," in *Deconstruction is/in America: A New Sense of the Political*, ed. Anselm Haverkamp (New York and London: New York University Press, 1997), 15, 24, 35.

connection of time to consciousness and the consequent relativity and subjectivity of time, its anti-linearity and multi-facetedness, contrary to the linearity and homogeneity of mechanical time, find their expression in Thomas Mann's *The Magic Mountain* (1924), Marcel Proust's *In Search of Lost Time* (1913-27) and stream-of-consciousness novels by Virginia Woolf and James Joyce who use the technique to reflect the meanderings and diversions of psychological time, as well as some postmodernist novels (e.g. Ian McEwan's *Atonement,* 2001). Paul Ricoeur has called Woolf's, Mann's and Proust's novels examples of *Zeitromanen* as they most fully reinvent the ways in which to illustrate the experience of private time.[13]

Late modernist and postmodernist notions of time

Late modernism and postmodernism conceptualize time differently from the old modernism. Literary fiction starts to reflect "a schismatic temporality" which is, in Frank Kermode's understanding, characterized by a new attitude to the past: while the old modernism finds a sense of order in the past, the newer modernism ignores it.[14] The 1980s brought the problematization of the notions of history and historiography, Francis Fukuyama's announcement of the end of history and Jean Francois Lyotard's rejection of grand narratives and their replacement with "little narratives." Whereas modernism questions and discards the past but does not express any skepticism towards the future, for many thinkers postmodernism does away with the notion of time itself.[15]

In the second half of the twentieth century, as Helga Nowotny argues, the notion of the future was substituted by the notion of the "extended present."[16] Fredric Jameson has declared "the end of temporality" and the collapse of human temporality into the "perpetual present": the "dramatic and alarming shrinkage of existential time and the reduction to a present that hardly qualifies as such any longer, given the virtual effacement of the past and the future that can alone define a present in the first place."[17] In Jameson's interpretation of Lacan, schizophrenia constitutes a breakdown of the connection between signifiers, which is related to the impossibility

[13] See Ricoeur, *Time and Narrative*, vol. 2, 112-125.

[14] Frank Kermode, *The Sense of An Ending. Studies in the Theory of Fiction with a New Epilogue* (Oxford: Oxford University Press, 2000), 115.

[15] Heise, *Chronoschisms. Time, Narrative and Postmodernism,* 35.

[16] See Helga Nowotny, *Time: The Modern and Postmodern Experience* (Cambridge: Polity Press, 1998).

[17] Fredric Jameson, "The End of Temporality," *Critical Inquiry* 29 (2003), 708.

to access the Symbolic. As a result, the individual lives in "a perpetual present with which the various moments of his or her past have little connection and for which there is no conceivable future on the horizon. … The schizophrenic does not know personal identity in our sense, since our feeling of identity depends on our sense of the persistence of the 'I' and the 'me' over time."[18]

For Zygmunt Bauman, the perpetual present of the western man, engaged in the pursuit of ephemeral fashions, trends and pseudoevents, created by the media, which by the overload of information facilitates forgetting, consists in a sequence of episodes and projects "hygienically insulated from their past as well as their future."[19] They do not add up to the sequential progress, contributing thus to the experience of "liquid times," which demand constant flexibility and adjustability.[20] As opposed to former realist literature, in literary fiction this schizophrenia or slicing of time is reflected by the formal fragmentation of the narrative. Time in such narratives is not presented as a sequence of moments, nor is it perceived as such by the subject who experiences "isolated, disconnected, discontinuous material signifiers which fail to link into a coherent sequence."[21]

On the one hand, therefore, the postmodernist fixation on the present is connected with the capitalist economy, but, on the other hand, it is linked with the temporal organization of the media; with the impact of technology, television and the Internet. The digital culture has blurred the boundary between public and private time. For Paul Virilio, "the new technological time has no relation to any calendar of events nor to any collective memory. It is pure computer time, and as such helps construct a permanent present, an unbounded timeless intensity."[22] Bauman also describes the liquidity of modernity as "unbounded time" because "fluids … neither fix space nor bind time."[23] "The insubstantial, instantaneous

[18] Fredric Jameson, "Postmodernism and Consumer Society," in *The Anti-Aesthetic: Essays on Postmodern Culture*, ed. Hal Foster (Port Townsend, WA: Bay Press, 1983), 119.
[19] Zygmunt Bauman, *Globalization. The Human Consequences* (New York: Columbia University Press, 1998), 88.
[20] See Zygmunt Bauman, *Liquid Modernity* (Cambridge: Polity Press, 2000), 91-129 and *Liquid Times. Living in an Age of Uncertainty* (Cambridge: Polity Press, 2007).
[21] Jameson, "Postmodernism and Consumer Society," 119.
[22] Paul Virilio, *The Lost Dimension* (Los Angeles: Semiotext(e), 1991), 15.
[23] Bauman, *Liquid Modernity*, 2.

time"[24] of the media allows immediate access to any temporal segment, disturbing thus the temporal order. The process of channel hopping and surfing a few websites at a time results in a juxtaposition of "radically disjunct times and spaces in visual simultaneity" and forces the viewer to concentrate "on a rapid succession of microevents," separated from the past or future.[25] Temporality becomes thus schizophrenic, or kaleidoscopic. It can be described by Allan Rodway's category of fugal time to point to narratives in which "time can be treated like space and shifted about kaleidoscopically; it can be layered, as in counterpoint." In fugal narratives "overall time-structuring is contrapuntal, but each separate episode, of course, is linear."[26]

An example of a fugal narrative is Penelope Lively's *Moon Tiger* (1987) in which she alternates times, tenses and standpoints: some of the novel is narrated in the first person in the past tense by Claudia, the protagonist, looking back on her life, while other passages are narrated in the present tense in the third person. Claudia's memories are alternated with the same events narrated from the other characters' points of view. The experience of time as something ordered and sequential is discarded here as well as the objectivity of reality and history: time is "broken up into a hundred juggled segments, each brilliant and self-contained so that hours are no longer linear but assorted like bright sweets in a jar."[27] She thus supports the kaleidoscopic view of time, comparing it to computer time:

> The question is, shall it or shall it not be linear history? I've always thought a kaleidoscopic view might be an interesting heresy. Shake the tube and see what comes out. Chronology irritates me. There is no chronology inside my head. I am composed of a myriad Claudias who spin and mix and part like sparks of sunlight on water. The pack of cards I carry around is forever shuffled and re-shuffled; there is no sequence, everything happens at once. The machines of the new technology, I understand, perform in much the same way: all knowledge is stored, to be summoned up at the flick of a key.[28]

Likewise, in Ian McEwan's *The Child in Time* (1987) the protagonist, Stephen, is "lost in time," arrested in a perpetual present when his

---

[24] Bauman, *Liquid Modernity*, 118.
[25] Heise, *Chronoschisms. Time, Narrative and Postmodernism,* 26.
[26] Allan Rodway, *The Truths of Fiction* (New York: Schoken Books, 1971), 117, 126.
[27] Penelope Lively, *Moon Tiger* (London: Penguin Books, 1988), 107, 108.
[28] Lively, *Moon Tiger*, 2.

daughter is kidnapped. His friend, Thelma, is a theoretical physicist and through her the insights of quantum physics are introduced. They explain a variety of experiences of time with its changing quality and speed Stephen undergoes: the repressive time of a government committee, the memories of his lost child, the empty timelessness of watching television, the aridity of eternal present, the peculiar slowing of time when he is on the verge of a car accident or the vision during which he can see his mother pregnant with him through a window pane.

Postmodern(ist) temporalities

As postmodernist literary fiction reflects time revealed by scientific discoveries, such as quantum mechanics, but also geology, evolution, radioactivity and the expansion of the universe, fictional time and temporality are no longer rooted in character psychology. The nanoseconds of computer time in James Gleick's "the epoch of nanosecond"[29] are juxtaposed with "deep time" (geological time) and cosmic time; the macroscopic scale of relativity theory with the microscopic scale of quantum mechanics, producing time dependent on the scale of the event. Our sense of time thus "'falls apart' into different temporal scales which have a simultaneous reality and yet cannot be contemplated simultaneously"[30] since there is no underlying "intrinsic" notion of time, clamping all the perspectives.[31]

Accordingly, postmodernist writers have responded to these disjointed multiple temporalities by producing narratives in which temporality is divorced from the human scale. If various points of view in a modernist novel complement each other to give a consistent account of the events, various perspectives in a postmodernist novel are contradictory and lack the organizing narrator.[32] While a modernist novel constitutes a juxtaposition of private times, which can be comprehended as a literary response to Einstein's relativity theory, I argue that the postmodernist novel with its mutually exclusive tales transpires to be the counterpart of Hugh Everett's many-universes interpretation of quantum mechanics. Modernist literature depicts thus human temporality whereas postmodernist literature reacts to and fictionalizes non-commonsensical theories of

---

[29] James Gleick, *Faster. The Acceleration of Just About Everything* (New York: Pantheon Books, 1999), 6.

[30] Heise, *Chronoschisms. Time, Narrative and Postmodernism,* 42.

[31] See Paul Davies, *About Time. Einstein's Unfinished Revolution* (London: Penguin Books, 1995), 178-182.

[32] Heise, *Chronoschisms. Time, Narrative and Postmodernism,* 53.

quantum physics, employing them as a narrative framework and/or thematizing them.

Physics fiction and Quantum fiction

The term "physics fiction" was used for the first time by Vladimir Nabokov in his *Ada, or Ardor: A Family Chronicle* (1969). There is some ambiguity in its employment by various critics and scholars. Alan J. Friedman includes in the category both fiction with direct references to physics and its metaphorical employment; "using models, images, and language from science to discuss an aspect of the human condition that the author perceives to have similarities to a topic in the physical universe."[33] Dennis Bohnenkamp, on the other hand, defines the strand in fiction as depending "quite heavily, and quite overtly" on the new physics; it is "a fiction quite literally 'about' physics."[34] He distinguishes physics fiction from science fiction which similarly relies on physics. In contrast, Amit and Maggie Goswami employ the term "physics fiction" to indicate the strand in science fiction which uses physical laws accurately.[35] I will use the term "physics fiction" to refer to narratives which relate to physics explicitly or implicitly, and "quantum fiction" to describe narratives which relate to quantum physics (explicitly or implicitly).

Quantum fiction, a term coined by the American novelist Vanna Bonta, and introduced with her book entitled *Flight: A Quantum Fiction Novel* (1995), refers to a new literary genre that appeared in the twentieth century. It reflects the experience of reality as described by quantum mechanics. Quantum fiction, "a new term bringing science up to speed, for when thought meets matter,"[36] constitutes, in Bonta's delineation, "the realm of all possibilities," as opposed to science fiction that depicts "the real."[37] Contrary to science fiction, primarily defined by content, quantum fiction is mainly identified by the technique of writing, and anything can become its subject matter; the narrative can but does not have to be

[33] Alan J. Friedman, "Contemporary American Physics Fiction," *American Journal of Physics* 47 no. 5 (1979), 392.
[34] Dennis Bohnenkamp, "Post-Einsteinian Physics and Literature: Toward a New Poetics," *Mosaic* 22 no. 3 (2002), 26.
[35] See Amit Goswami with Maggie Goswami, *The Cosmic Dancers: Exploring the Physics of Science Fiction* (New York: Harper, 1983).
[36] Vanna Bonta, *Flight: A Quantum Fiction Novel* (San Diego: Meridian House, 1997), 339.
[37] "Vanna Bonta Talks About Quantum Fiction," author interview by Laurel van der Linde, 2007, <IMDB.com>.

science-oriented. The narratives employ non-chronological plots and disrupted temporalities to render the characters' experience of quantum reality, such as parallel worldlines in multiverse, synchronicity (mirroring quantum entanglement), the impossibility of (quantum) suicide, consciousness as an active participant in the creation of reality, the unpredictability of reality, holism, and so on.

A Spanish author, Gregorio Morales, enumerates the features of quantum aesthetics in his essay "El cadaver de Balzac" ("Balzac's Corpse," 1998), a manifesto of the movement called Quantum Aesthetics. Some of the features are: "1) The artist has the power to break away from the walls imposed by Newton. 2) Reality is not restricted to what is visible or knowable. 3) The implicate and invisible order is differentiated from the explicate and manifest order. 4) Matter and spirit are the same thing; they form a continuum. 5) The laws of causality become relative. In fact, we can even find an acausal relationship between the psychological and the physical realms – a synchronicity. 6) Nothing is distinct. Even the most distant corpuscles influence one another. 7) Both observer and observed are tied together. … 8) We can remain within the visible world, that is, the preceding tenets should not be taken as fetishes. … 10) a primary concern for freedom and diversity" which supports "an ethics based on tolerance, anti-dogmatism, and a strong defense of democracy. … 11) We prefer the extraordinary rather than the common, eroticism rather than the physiology of sex, maturity rather than juvenility, individuality rather than collectivity, internal transformation rather than surgery, and consciousness rather than fluids."[38]

The term "quantum fiction" seems to be the counterpart of Susan Strehle's term "actualism" which she derives from Werner Heisenberg. He distinguishes between the real and the actual – reality is active, dynamic, fluent, provisional, actual, rather than real at the subatomic level. Strehle employs the term "actualistic fiction" to refer to a literary representation of reality as understood by the new physics.[39] She rejects the division of fiction into realistic, reflecting Newtonian universe, and anti-realistic, interrogating the reality of the external world. Although contemporary fiction moves away from realism, Strehle argues, it does not abandon its interest in reality, but only transforms the way in which reality is understood. Both actualistic fiction and the new physics render reality as

[38] After Francisco Javier Peñas-Bermejo, "Quantum Literature," in *The World of Quantum Culture,* eds. Manuel J. Caro, John W. Murphy (Westport, Connecticut, London: Praeger, 2002), 61 n. 17.
[39] Susan Strehle, *Fiction In the Quantum Universe* (Chapel Hill and London: The University of North Carolina Press, 1992), 7.

relative, irregular, subjective, statistical, uncertain and energetic,[40] and propose consciousness to be an aspect of reality.

The crucial difference between modernist and postmodernist fiction, according to Strehle, is that modernist fiction mostly depicts the Newtonian view of the world, while postmodernist fiction reflects the world as proposed by the new physics. Both modernist and postmodernist fictions are non-chronological, offering new paradigms of causality and "simultaneity, reflexive reference, and juxtaposed motionless glimpses fit into a formal unity," yet, in modernist literature "the picture of reality emphasizes its own composed nature … but it composes according to Newtonian principles."[41] A perfect example is Durrell's *Alexandria Quartet* where he misinterprets Einstein's theories, confusing them with quantum mechanics.[42] Although the *Quartet* and other modernist novels aspire to mirror the reality of the new physics, they do not expose the plurality of subjective truths or the modernist conviction of the unknowability of truth, but allow the binding of the various perspectives together, presenting thus the ultimate truth.[43] A 'proper' example of relativity theory in fiction is a work by an American writer, Edmund Wilson, entitled *I Thought of Daisy* (1929), where Daisy is shown from a few perspectives, mutually exclusive and contradictory.

A Guyanese writer, Wilson Harris, is one of the writers who have most profoundly contributed to the poetics of quantum fiction. He himself describes his writings as quantum fiction and emphasizes the fact that he was not familiar with the concepts of quantum physics at the time of writing *Palace of the Peacock* (1960).[44] His narrative structure is fluent and multiple, informed by wave-particle duality and multiverse theory. His technique consists in fluid metamorphosing imagery, "the convertible imageries,"[45] representing simultaneous parallel possibilities, described by

[40] Strehle, *Fiction In the Quantum Universe*, 3, 6, 15.

[41] Strehle, *Fiction In the Quantum Universe*, 17.

[42] See Lawrence Durrell's "Preface" in *Balthazar* (Dutton: New York, 1958) and "Space, Time and Poetry" in *A Key to Modern British Poetry* (Norman: University of Oklahoma, 1970), 24-48.

[43] See Sarah Cain, "The Metaphorical Field: Post-Newtonian Physics and Modernist Literature," *The Cambridge Quarterly*, vol. 28, no. 1 (1999), 54-55.

[44] See "Michael Gilkes Interviews Sir Wilson Harris," *Knews* (July 18 2010), <http://www.kaieteurnewsonline.com/2010/07/18/michael-gilkes-interviews-sir-wilson-harris-2/>.

[45] Wilson Harris, "The Author's Note," in *Palace of the Peacock* (London, Boston: Faber and Faber, 1988), 10.

Nick Herbert as "polyhistory."[46] Harris describes the difference between traditional and quantum fiction thusly: "The quantum concept is that if one fires out an object, it breaks into particles and waves. Conventional novelists go along a linear road, but the quantum split can bring the past into the present in a new art of fiction."[47]

What is interesting, while scientists leave Einstein's theories in the macroscopic scale and quantum mechanics in the microscopic one, both separated from everyday reality, fiction writers adjust them to the human scale and effect their import on the everyday reality of the characters. The notions of the new physics referring to time and temporality that writers rework aesthetically include, apart from relativity theory, parallel universes, the backward flow of time and holism.

Parallel universes

According to Everett's multiverse theory, the wavefunction does not collapse, as in Copenhagen's interpretation of quantum mechanics, but reality becomes a many-branched tree, wherein every possible outcome is realized in the parallel actual universes. In fiction the first forking-path narrative was D. R. Daniels's "The Branches of Time" in which in 1935, twenty two years before Everett's many-worlds theory, he utilizes quantum mechanics for a time travel plot. Travelling to the past leads here to producing alternative universes, which exposes the attempts to alter the past as futile.[48] However, it is not Daniels's short story but the more influential "The Garden of Forking Paths" (1941) by Jorge Luis Borges that is usually mentioned as the first text illustrating parallel time streams. With this story, Borges invents a hypertext novel, but he also describes a theory of the universe based upon the structure of such a novel. Forking-

---

[46] Nick Herbert, *Quantum Reality* (New York: Anchor Press/Doubleday, 1985), 27. Rebekka Edlung borrows the term "polyhistory" to analyse Harris's fiction. See her "Carnival and quantum theory. Metaphors of identity in Wilson Harris's The Carnival Trilogy," in *The Society for Caribbean Studies Annual Conference Papers*, vol. 7, ed. Sandra Courtman (2006), 1-7.

[47] Maya Jaggi, "Redemption Song," *The Guardian* (Saturday 16 December 2006), <http://www.theguardian.com/books/2006/dec/16/featuresreviews.guardianreview15>.

[48] Although time-travel is allowed by the general theory of relativity, even if we had a time machine, travelling in time would mean producing a new branch, rather than taking part in two time-lines in one universe contemporaneously. Previous time travel stories, pioneered by H. G. Wells's *The Time Machine* (1895), utilize relativity theory as the underlying scientific framework.

path narratives encompass the narrative of alternate history (multiverse or a history counterfactual to ours, e.g. Jasper Fforde's *The Eyre Affair*, 2001), the narrative of transworld exploration (where all the worlds are located within the same space-time continuum, e.g. *Star Trek*, 1966-) and the time-travel one (where the protagonist triggers the development of new branches by travelling in time).[49]

Everett's theory presupposes no spatiotemporal relations between the worlds – the universe and the observer bifurcate into many different incarnations but in each of them s/he observes just one definite state. This rule is usually violated by literary fiction, e.g. in Jeanette Winterson's *Sexing the Cherry* (1989) or Andrew Crumey's *Sputnik Caledonia* (2008). In *Sexing the Cherry* two protagonists, Jordan and the Dog-Woman, exist in the seventeenth and twentieth centuries simultaneously. Whereas the bodies are different, the consciousness, traits of character and attitudes are the same. The characters do not have epistemic access to parallel trajectories; sometimes however, they have flashes from the other life.[50] Winterson thereby utilizes the new physics to underline postmodern multi-facetedness and the fragmentation of the subject.

The Scottish theoretical physicist and writer Andrew Crumey has employed the framework of parallel universes in all of his novels. In the first part of his *Sputnik Caledonia*, a nine-year-old boy living in Kenzie, Scotland, dreams of being a cosmonaut and asks "a disembodied transcendental higher intelligence"[51] to take him to another world. In the second part, Robert, now at the age of nineteen, is trained to go on a space mission in the Installation, the British Democratic Republic's top secret military base. During the mission he is to free-fall from a rocket so as to come into telepathic contact with a black hole, called the Red Star. He dies but the voice informs him, "*Your life is over, Robert – but not your story.*"[52] In part III, back in Kenzie, a forty-year-old Robert Coyle who claims to have seen "many worlds"[53] appears in the park. It transpires that the boy who dreamt of being a cosmonaut died at the age of nineteen. The three parts can thus be treated as parallel lives of the protagonist as the text is full of intrusions from his other trajectories. Although the novel seems

[49] Marie-Laure Ryan, "From Parallel Universes to Possible Worlds: Ontological Pluralism in Physics, Narratology and Narrative," *Poetics Today* 27:4 (2006), 656-658.

[50] See Sonia Front, *Transgressing Boundaries in Jeanette Winterson's Fiction* (Frankfurt am Main: Peter Lang, 2009), 182-186.

[51] Andrew Crumey, *Sputnik Caledonia* (London: Picador, 2008), 121.

[52] Crumey, *Sputnik Caledonia,* 422. Emphasis in original.

[53] Crumey, *Sputnik Caledonia*, 452.

to be written in a traditional chronological way, the vision of reality it presents is far from classical. When one gets the holistic view of the three parts of the book, they do not coalesce into a linear narrative but a bifurcating one.

The arrow of time

Other narratives which utilize the exploding temporalities of the new physics embrace those which violate the direction of the arrow of time. This includes texts in which the character's biological processes are reversed and s/he lives towards birth, and/or narratives with reversed chronology. While physics does not differentiate between the forward and backward directions of time, as Stephen Hawking explicates, the phenomenon of the effect following the cause can be attributed to the thermodynamic arrow of time, the result of the second law of thermodynamics, according to which in an isolated system entropy tends to increase with time.[54] This is accompanied by the psychological arrow of time since we have the sense of continuing from the past into the future. Stuart Hameroff maintains that the second law of thermodynamics does not apply "in quantum systems which may become more ordered."[55] Time can flow backwards here. Information can be sent backward in time, for example by OR events (Roger Penrose's name for conscious moments), which is used e.g. in quantum cryptography.[56] Whether the reversal of time's arrow is possible in physics, or not, certainly it has captured the imagination of several writers. The arrow of time is disrupted, for instance, in Holocaust fiction where the irrational temporality of the novel is an attempt to deal with an irrational unrepresentable experience.

In Martin Amis's *Time's Arrow* (1991) the narrator, a kind of secondary consciousness inhabiting a German doctor, experiences time running backwards until the protagonist disappears in his mother's womb. Whilst with the reversed chronology the effects precede the cause and the narrator is confused with the world around him – in which people vomit

---

[54] Stephen Hawking, *A Brief History of Time: From the Big Bang to the Black Holes* (London: Bantam, 1988), 160. See also Peter Conveney, Roger Highfield, *The Arrow of Time. The Quest to Solve Science's Greatest Mystery* (London: Flamingo, 1991), 147-181.

[55] Stuart Hameroff, "Time, Consciousness and Quantum Events in Fundamental Spacetime Geometry," *NATO Science Series II: Mathematics, Physics and Chemistry*, vol. 95 (2003), <http://www.quantumconsciousness.org /time.htm>.

[56] Hameroff, "Time, Consciousness and Quantum Events in Fundamental Spacetime Geometry."

their food onto plates, form vegetables and meat and then uncook them to sell them in the supermarket, or after seeing the doctor patients wait for an hour in the waiting room – finally in Auschwitz things start to make sense: "Our preternatural purpose? To dream a race. To make a people from the weather. From thunder and from lightning. With gas, with electricity, with shit, with fire."[57] Inverted temporality reflects thus the Nazi inverted ways of thinking and reasoning.

In *The White Hotel* (1981) D.M. Thomas in six sections with different narrators employs the technique of disjunction, deploying a collection of documents: a poem, prose, a journal, a psychoanalytic case study and postcards to slowly arrive at the massacre in Babi Yar, Kiev. The protagonist is Sigmund Freud's patient, Anna G., who is psychoanalyzed as she suffers from acute psychosomatic pains. While Freud looks for the justification in her childhood, in fact this is only part of the story. Anna's pains appear to be the memories of the future since she is going to be violated in Babi Yar. Freud's method appears thus a consistent but partial form of rational control over human psyche; he assumes human identity to be essential and coherent and does not take Anna's subjection to the violence of history into account. The ruptured temporalities of *Time's Arrow* and *The White Hotel* corroborate that genocide and suffering resist all attempts to explain them theoretically; "you can no more imagine Babi Yar than you can conceive a world where time and causality run in opposite directions."[58]

Holism

Holism is the idea that natural systems and their properties should be viewed as wholes, rather than the collections of parts, which results from the assumption that systems operate as wholes and that their workings cannot be fully understood merely on the basis of their constituent parts. Ontological holism was advocated by two independent thinkers simultaneously: the physicist David Bohm in his *Wholeness and the Implicate Order* (1980) and the neurophysiologist Karl Pribram. Bohm elucidates: "Parts are seen to be in immediate connection, in which their dynamical relationships depend, in an irreducible way, on the state of the whole system … . Thus one is led to a new notion of unbroken wholeness

[57] Martin Amis, *Time's Arrow* (Vintage, 1992), 119-120.
[58] Marie-Laure Ryan, "Temporal Paradoxes in Narrative," *Style*, vol. 43, no. 2 (2009), 156.

which denies the classical idea of analyzability of the world into separately and independently existing parts."[59]

It must be noted, however, that popularizers of quantum physics overemphasize the significance of human consciousness in quantum theory while in fact few physicists support this interpretation (Von Neumann-Wigner interpretation). In spite of that, the decisive role of consciousness is the central point of many popular science books and "paranormalists of every stripe, from psychic to astrologers to physicists to cosmologists, [who] proclaim the oneness of the human mind and the fabric of the cosmos."[60] Similarly, Bohm's model of wholeness produced widespread interest in religious, philosophical and New Age circles, and his ideas were used in the holographic model of consciousness.

The idealist interpretation transpires to be the theme of many fictional works. It might be so because fiction writers rely heavily on popular science books, and have been influenced by their stress on quantum consciousness, holism and unbroken wholeness, or perhaps because these ideas seem to have an attractive literary potential. Therefore, holistic imagery, mainly rejected by 'serious' science, abounds in literary fiction, entering common imagination through this backdoor. In the holistic unity, figured as quantum/universal consciousness, the temporal dimensions break down. Already T.S. Eliot in his *Four Quartets* (1945) advocates holism and the simultaneity of past, present and future, proposed by Einstein: "the end and the beginning were always there/ Before the beginning and after the end./ And all is always now." In the aforementioned *Sputnik Caledonia* the Red Star is the embodiment of the idea of holism as it stands for universal consciousness. By means of the Star's mind the protagonist connects with his own consciousness in parallel universes. The Star appears thus as a kind of channel to bifurcating universes but contemporaneously, it constitutes a repository of time, history, memory and selves; a totality of contingent pasts and futures, existing all-together.

In Scarlett Thomas's *The End of Mr. Y* (2006), the protagonist, Ariel Manto is able to mentally access a kind of virtual reality called Troposphere where she can leap into other people's minds. Troposphere,

---

[59] David J. Bohm and Basil J. Hiley, "On the Intuitive Understanding of Nonlocality as Implied by Quantum Theory," in *Foundations in Physics,* (Princeton: Princeton University Press, 1961), 319. See also David Bohm, *Wholeness and the Implicate Order* (London: Routledge, 1980).

[60] Victor J. Stenger after Elizabeth Leane, "Knowing Quanta: The Ambiguous Metaphors of Popular Physics," *The Review of English Studies*, New Series, vol. 52, no. 207 (2001), 422.

as Thomas presents it, is universal consciousness, the underlying reality, where the distinctions between temporal dimensions and between mind and matter collapse, and all things are connected. In her attempt to theorize the experience of Troposphere Ariel argues,

> we all potentially know everything about what other people are thinking and doing. We all potentially have access to one another's thoughts. I wonder properly where the Troposphere is, and what it is, now that I'm convinced it isn't just a figment of my imagination. Is it hovering less than a particle away from us, perhaps in another dimension to which we have access only some of the time? Or does it work in another way entirely? But I am suddenly sure that the moment when you catch someone's eye, or the moment when you think someone's looking at you, or the moment when you think of someone and then they ring, or the moment when you start getting lost in a building you know so well because most other people in it are lost – these aren't accidents. They relate in some way to the structure of the physical world, to the fact that all our minds are as connected as everything else.[61]

At the end of the book when Ariel and her boyfriend Adam decide to stay in Troposphere, their earthly bodies already dead, they find themselves in a quantum vacuum, that is "the basic, fundamental and underlying reality of which everything in this universe – including ourselves – is an expression."[62] The vacuum is conceptualized as "a sea of potential," where "*You now have infinite choice.*"[63]

Holistic is also the vision of the universe proposed by Jeanette Winterson in *Gut Symmetries* (1997). For her, the universe and human beings are "patterned together."[64] Identity is in the perpetual state of becoming, its constituents fluctuate to expose in the holographic manner "The dead, time, light patterns of millennia opening in your gut."[65] "Gut" and "gut feelings" are paralleled here with GUT, an attempt at a single theory to explain the world. One of those theories, constantly referred to in the novel, is superstring theory, which proffers the particles and fundamental forces of nature to be the vibrations of super-symmetric strings. By means of the images of guts but also of a river, time is

---

[61] Scarlett Thomas, *The End of Mr. Y* (Edinburgh, New York, Melbourne: Canongate, 2007), 380.

[62] Danah Zohar in collaboration with I.N. Marshall, *The Quantum Self. Human Nature and Consciousness Defined by the New Physics* (New York: William Morrow and Company, Inc., 1990), 225.

[63] Thomas, *The End of Mr. Y*, 452.

[64] Jeanette Winterson, *Gut Symmetries* (London: Granta, 1997), 98.

[65] Winterson, *Gut Symmetries*, 218.

conceptualized as a vortex. Rather than referring to the simplistic understanding of the movement of the river as an unvarying linear flow of water, Winterson encompasses its turbulences, twists and countercurrents: "A river cannot flow against its current, but it can flow in circles: its eddies and whirlpools regularly break up its strong press forward. The riverrun is maverick, there is a high chance of cross-current, a snag of time that returns us without warning to a place we thought we had sailed through long since."[66] This model of time presupposes polytemporal experience, fluctuating modes of the past, present and future, memories spiraling back to the past, the labyrinthine quality of mental life. Similarly to Michel Serres's understanding, time flows in a chaotic way; it "percolates."[67]

Conclusion

Quantum fiction manifests itself as a response to the crisis of representation; to the need, recognized by many authors and critics, for creating a new language, literary technique and genre to describe quantum reality. Our language, which relies on commonsense logic is not capable of dealing with the issues of physics and the new metaphysics. McEwan argues that "We are forced to imagine something beyond language."[68] Hayles writes about the difficulties of language to reflect new notions of time, especially when the relationship between cause and effect is multidirectional, "when every cause is simultaneously an effect, and every effect is also a cause – the language of cause and effect is inadequate to convey the mutuality of the interaction. Causal descriptions will not do because causal terminology implies a one-way interaction that falsifies the essence of what we want to convey."[69] Therefore, as I have demonstrated, various writers have developed new ways of addressing the concepts of temporality, from relativity theory, to superstring theory, to the arrow of time, to holism, to parallel universes (whether interpreted as the enactment of Everett's theory, or bifurcating narratives, or hypertext, or polyhistory

---

[66] Winterson, *Gut Symmetries*, 104.

[67] See Michel Serres with Bruno Latour, *Conversations on Science, Culture, and Time*, trans. Roxanne Lapidus (Ann Arbor: University of Michigan Press, 1995), 57-59.

[68] Ian McEwan, Ryan Roberts, *Conversations with Ian McEwan* (Jackson: University Press of Mississippi, 2010), 141.

[69] Hayles, *The Cosmic Web. Scientific Field Models and Literary Strategies in the Twentieth Century*, 20.

or Multiple Drafts novels[70]). Through representing these concepts, fiction attempts to address the questions of relativity and the subjectivity of time, the temporality of psyche, and, most important of all, the relation between time and human identity.

References

Amis, M., 1992, *Time's Arrow*, New York: Vintage.

Bakhtin, M. M., 1981, "Forms of Time and of the Chronotope in the Novel," in Holquist, M., ed., Emerson C., and Holquist, M., trans., *The Dialogic Imagination: Four Essays by M. M. Bakhtin,* Austin: University of Texas Press, 84-258.

Bauman, Z., 1998, *Globalization. The Human Consequences*, New York: Columbia University Press.

—. 2012, *Liquid Modernity*, Cambridge: Polity Press.

—. 2007, *Liquid Times. Living in an Age of Uncertainty*, Cambridge: Polity Press.

Bohm, D., 1980, *Wholeness and the Implicate Order*, London: Routledge.

Bohm, D. J., and Hiley, B. J., 1975, "On the Intuitive Understanding of Nonlocality as Implied by Quantum Theory," *Foundations of Physics* 5(1), 93-109.

Bohnenkamp, D., 2002, "Post-Einsteinian Physics and Literature: Toward a New Poetics," *Mosaic* 22 (3), 19-30.

Bonta, V., 1997, *Flight: A Quantum Fiction Novel*, San Diego: Meridian House.

Burn, S. J., 2012, "Reading the Multiple Drafts Novel," *Modern Fiction Studies* 58 (3), 436-458.

Cain, S., 1999, "The Metaphorical Field: Post-Newtonian Physics and Modernist Literature," *The Cambridge Quarterly* 28 (1), 46-64.

Conveney, P., and Highfield, R., 1991, *The Arrow of Time. The Quest to Solve Science's Greatest Mystery*, London: Flamingo.

Crumey, A., 2008, *Sputnik Caledonia*, London: Picador.

Davies, P., 1995, *About Time. Einstein's Unfinished Revolution*, London: Penguin Books.

Durrell, L., 1958, Preface to *Balthazar,* Dutton: New York.

—. 1970, "Space, Time and Poetry," in *A Key to Modern British Poetry.* Norman: University of Oklahoma, 24-48.

[70] See Stephen J. Burn "Reading the Multiple Drafts Novel," *Modern Fiction Studies*, vol. 58, no. 3 (2012), 436-458 where he interprets a few novels through the lens of Daniel Dennett's the Multiple Drafts model of consciousness.

Edlung, R., 2006, “Carnival and quantum theory. Metaphors of identity in Wilson Harris’s The Carnival Trilogy,” in Courtman, S., ed., *The Society for Caribbean Studies Annual Conference Papers*, Vol. 7, 1-7.

Friedman, A. J., 1979, “Contemporary American Physics Fiction,” *American Journal of Physics* 47 (5), 392–395.

Front, S., 2009, *Transgressing Boundaries in Jeanette Winterson’s Fiction*, Frankfurt am Main: Peter Lang.

Gilkes, M., 2010, “Michael Gilkes Interviews Sir Wilson Harris,” *Knews*, July 18 2010, <http://www.kaieteurnewsonline.com/2010/07/18/michael-gilkes-interviews-sir-wilson-harris-2/>.

Gleick, J., 1999, *Faster. The Acceleration of Just About Everything*, New York: Pantheon Books.

Goswami A., and Goswami, M., 1983, *The Cosmic Dancers: Exploring the Physics of Science Fiction*, New York: Harper.

Hameroff, S., 2003, “Time, Consciousness and Quantum Events in Fundamental Spacetime Geometry,” *NATO Science Series II: Mathematics, Physics and Chemistry*, <http://www.quantumconsciousness.org/time.htm>.

Harris, W., 1988, *Palace of the Peacock*, London, Boston: Faber and Faber.

Hawking, S., 1988, *A Brief History of Time: From the Big Bang to the Black Holes*, London: Bantam.

Hayles, K. N., 1984, *The Cosmic Web. Scientific Field Models and Literary Strategies in the Twentieth Century*, Ithaca and London: Cornell University Press.

Heise, U.K., 1997, *Chronoschisms. Time, Narrative and Postmodernism*, Cambridge: Cambridge University Press.

Herbert, N., 1985, *Quantum Reality*, New York: Anchor Press/Doubleday.

Jaggi, M., 2006, “Redemption Song,” *The Guardian*, <http://www.theguardian.com/books/2006/dec/16/featuresreviews.guardianreview15>.

Jameson, F., 2003, “The End of Temporality,” *Critical Inquiry* 29, 695-718.

—. 1983. “Postmodernism and Consumer Society,” in Foster, H., ed., *The Anti-Aesthetic: Essays on Postmodern Culture*, Seattle: Bay Press, 11-125.

Joyce, J., 2013, *Ulysses*, London: Urban Romantics.

Kermode, F., 2000, *The Sense of An Ending. Studies in the Theory of Fiction with a New Epilogue*, Oxford: Oxford University Press.

Kern, S., 2003, *The Culture of Time and Space, 1880-1918: With a New Preface*, Cambridge, Massachusetts, London, England: Harvard University Press.

Kuhn, T., 1996, *The Structure of Scientific Revolutions*, Chicago: University of Chicago Press.

Leane, E., 2001, "Knowing Quanta: The Ambiguous Metaphors of Popular Physics," *The Review of English Studies*, New Series, 52 (207), 411-431.

Lewis, W., 1957, *Time and Western Man*, Boston: Beacon Press.

Linde, L., 2007, "An Interview With Author, Vanna Bonta Regarding *Flight, A Quantum Fiction Novel*," <http://www.gather.com/viewArticle. action?articleId=281474977238495>.

Lively, P., 1988, *Moon Tiger*, London: Penguin Books.

McEwan, I., 2010, *The Child in Time*, Random House.

McEwan, I., Roberts, R., 2010, *Conversations with Ian McEwan,* Jackson: University Press of Mississippi.

Nowotny, H., 1998, *Time: The Modern and Postmodern Experience.* Cambridge: Polity Press.

Parker, J. A., 2007, *Narrative Form and Chaos Theory in Sterne, Proust, Woolf and Faulkne*r, NY: Palgrave Macmillan.

Peńas-Bermejo, F. J., 2002, "Quantum Literature," in Caro, M. J., Murphy, J. W., eds., *The World of Quantum Culture*, Westport, Connecticut, London: Praeger, 47-70.

Ricoeur, P., 1990, *Time and Narrative,* Volume 2, McLaughlin, K., and Pellauer, D., London, trans., Chicago: University of Chicago Press.

Rodway, A., 1971, *The Truths of Fiction*, New York: Schoken Books.

Ryan, M., 2006, "From Parallel Universes to Possible Worlds: Ontological Pluralism in Physics, Narratology and Narrative," *Poetics Today* 27 (4), 633-674.

—. 2009, "Temporal Paradoxes in Narrative," *Style*, 43 (2), 142-163.

Serres, M., and Latour, B., 1995, *Conversations on Science, Culture, and Time*, Lapidus. R., trans., Ann Arbor: University of Michigan Press.

Shakespeare, W., 2003, *Hamlet*, Irvine, CA: Saddleback Educational Publ.

Strehle, S., 1992, *Fiction In the Quantum Universe*, Chapel Hill and London: The University of North Carolina Press.

Thomas, D.M., 2012, *The White Hotel*, Hachette UK.

Thomas, S., 2007, *The End of Mr. Y*, Edinburgh, New York, Melbourne: Canongate.

Urry, J., 2000, *Sociology Beyond Societies: Mobilities for the Twenty-First Century*, New York: Routledge.

Virilio,P., 1991, *The Lost Dimension*, Los Angeles: Semiotext(e).

West-Pavlov, R., 2013, *Temporalities*, New York: Routledge.
Whitrow, G.J., 1989, *Time in History. Views of Time from Prehistory to the Present Day*, Oxford, N.Y.: Oxford University Press.
Winterson, J., 1997, *Gut Symmetries*, London: Granta.
—. 1989, *Sexing the Cherry*, London: Vintage.
Woolf, V., 1996, *Mrs. Dalloway*, Ware: Wordsworth Editions.
Zohar, D. in collaboration with Marshall, I. N., 1990, *The Quantum Self. Human Nature and Consciousness Defined by the New Physics*, New York: William Morrow and Company, Inc.

CHAPTER THREE

THE TEMPORAL CITY: SPECTRAL TOPOGRAPHIES AND THE EPHEMERAL PERFORMANCE OF BANGKOK

KATARZYNA ANCUTA

In 1984, Michel de Certeau invited us to experience the post-modern city through walking, the activity he saw as a "space of enunciation," or "a spatial acting-out of the place."[1] This addition of the vector of movement to the otherwise static topography of the city has allowed us to shift the urbanistic discourse from conceptualizing the city as a fixed place to the exploration of urban practices, or what de Certeau referred to as "an immense social experience of lacking a place."[2] But the consequences of this act do not stop here. Since movement implies transition not only through space but also through time, it ultimately results in the creation of temporary cities, experienced in their momentary presence. De Certeau compared the practice of walking to performative speech acts, utterances not only describing what is happening but actively changing the reality they are describing. Walking, thus, becomes equivalent to performing the city through movement and associated with it incessant imposition of memories of the city's past (distant and near) onto the city's present.

Walking, however, is not the only form of movement encountered in the city. If we step down from our vantage point as a lonely voyeur, we immerse ourselves in the multiplicity of everyday urban practices through which the city is continuously being realized, performed one act at a time by its inhabitants. The city, a space filled with movement rather than a place filled with immovable structures or a static landscape, is characterized

[1] Michel de Certeau, *The Practice of Everyday Life*, trans. Steven Rendall (Berkley: University of California Press, 1997), 98.
[2] Certeau, *The Practice of Everyday Life,* 103.

by its impermeability, existing from one moment to another. This, however, is not a new sentiment. Already in 1925, Robert Park convincingly argued:

> The city is, rather, a state of mind, a body of customs and traditions, and the organized attitudes and sentiments that inhere in these customs and are transmitted with this tradition. The city is not, in other words, merely a physical mechanism and an artificial construction. It is involved in the vital processes of the people who compose it; it is a product of nature, and particularly a human nature.[3]

In 1937, Lewis Mumford saw the city as "a theater of social action"[4] and described it as "a special framework directed toward the creation of differentiated opportunities for a common life and a significant collective drama."[5] Drawing on Charles Landry's notion of "experience economy" bringing together everyday life, consumption and spectacle, as well as performance studies theories, in his essay on "Performing the City," Paul Makeham shifts the focus to the city's inhabitants who function both as actors and audience, breathing life into the theatre of the city through their collective actions, sign exchanges, story-telling, display and spectatorship, dialogue and conflict,[6] setting the stage for the ritual performance of their collective memory, imagination, hopes and dreams.[7] Makeham returns to Jonathan Raban's concept of the "soft city" and agrees with David Nentwick that

> [t]he city as we imagine it, the soft city of illusion, myth, aspiration, nightmare, is as real, maybe more real, than the hard city one can locate on maps, in statistics, in monographs on urban sociology and demography and architecture.[8]

Taking the argument even further, Steve Pile offers a Freudian reading of "real cities," insisting that the imaginary, the fantastic and the emotional must be accounted for in the analysis of modern city life. Pile introduces the concept of "phantasmagoria" into the discussion of cities and urban

---

[3] Robert E. Park, "Suggestions for the Investigation of Human Behaviour in the Urban Environment," in *Urban Culture: Critical Concepts in History and Cultural Studies*, Vol. 1, ed. Chris Jenks (London & NY: Routledge, 2004), 19.

[4] Lewis Mumford, "What is a City," in *The City Reader*, ed. Richard T. LeGates and Frederic Stout (London & NY: Routledge, 2009), 87.

[5] Mumford, "What is a City," 87.

[6] Paul Makeham, "Performing the City," *Theatre Research International* 30.2 (2005), <http://eprints.qut.edu.au/7099/1/7099.pdf>, 158.

[7] Makeham, "Performing the City," 152.

[8] David Nentwick quoted in Makeham, "Performing the City," 154.

practices. For Pile, phantasmagoria describes "an experience of movement, or a procession of things before the eyes" invoking the experience of things "beyond the immediately visible or tangible" and suggesting a "ghost-like" or "dream-like" quality of life[9] in space and time. Pile's richly psychoanalytical readings of the modern city, was inspired by Guy Debord's notion of psychogeography, which is a mode of observation designed to map people's experiences of the city achieved through a deliberate attempt to reimagine spaces and spatiality,[10] result in his identification of four elements of the phantasmagorias of the city life which bring its emotional quality to the forefront: dreams, magic, vampires and ghosts. The dreams of the city are mostly those of a bright future, speaking both of its desires and fears. City magic, steeped in rituals, works to make wishes come true in city life. Vampiric cities do their blood-work, exposing the pleasure and horror of life, death and sex in the city. Last but not least, ghosts linger over city places making visible its connection to memory and grief.[11]

Time in Thailand is usually discussed through the prism of anthropology, greatly influenced by E.T. Hall's theory of chronemics.[12] Thai culture, according to his model, characterized by its polychromic time orientation, is thus frequently described as unfocused and favouring broad goals and interpersonal relationships over promptness, efficiency and timeliness. The phrase "Thai time" is commonly used to account for the tardiness, procrastination, and general disregard for punctuality. In Bangkok, time is often said to stand still, frozen in a paralyzing encounter with the all-encompassing, never-ending traffic jam. This chapter approaches Bangkok time differently. In line with the studies of city life that draw attention to the everyday experience, urban practices, movement, memory and performing the city, this paper engages in a discussion of temporal cities realized through the symbolic, ritualistic performance of their inhabitants and manifestations of collective memory on the example of Bangkok, a mega-metropolis known for its rather precarious relationship with reality. Though all four of Pile's phantasmagorias can be successfully applied to a reimagining of Bangkok, this chapter will focus

---

[9] Steve Pile, *Real Cities: Modernity, Space and the Phantasmagorias of City Life* (London: Sage, 2005), 3.

[10] Pile, *Real Cities: Modernity, Space and the Phantasmagorias of City Life,* 12.

[11] Pile, *Real Cities: Modernity, Space and the Phantasmagorias of City Life* ,166-172.

[12] Edward T. Hall, "Monochromic and Polychromic Time," in *Intercultural Communication: A Reader*, eds. L. A. Samovar and R. E. Porter (Belmont CA: Wadsworth/Thomson Learning, 2003), 262-268.

on the two most obvious categories – ghosts and magic, exploring their connection to the city's past and present and the lingering temporalities they leave behind.

From Stage to Performance

"The key characteristic of Bangkok is chaos,"[13] begins Ross King in his *Reading Bangkok.* King sees the chaotic dimension of Bangkok mostly in terms of a spatial category, realised through multiple juxtapositions of dissimilar or incompatible architectural structures, and extreme layering achieved through the superimposition of physical and symbolic screens hiding the spaces of the city and their production behind them. He observes that chaos is frequently embraced by Bangkokians as a medium of survival, if not open resistance.[14] King concludes: "Both visually and in the sheer complexity of the messages (information) that the vision presents, Bangkok space is disordered, muddled, chaotic, a shambles, which is not to deny the seductive attraction of the muddle."[15]

Recognizing disorder as the underlying organising principle behind Bangkok's city planning (or lack of thereof), we are not surprised to see the city's consistent resistance to systematic cartographic representation. "[T]here is no map of Bangkok. There are only maps of Bangkok's streets,"[16] declares Richard O'Connor in his article on the Thai image of the city that rejects the flat, simultaneous physicality of the map.[17] O'Connor sees Bangkok as "a patchwork of named places,"[18] adding somewhat cryptically that "Bangkok is lived more like a poem is heard – a line at a time."[19] This last remark is significant as it highlights the temporal quality of the experience and its performative dimension. While a printed version of a poem can be accessed at will and in any order preferred by the reader, the recited poem comes to life for a brief moment in an act of its enunciation; it is acted out in space and time, and then it is gone forever. Each repeated performance is slightly different, as it unfolds under different circumstances. Each spectacle is intrinsically ephemeral, as

[13] Ross King, *Reading Bangkok* (Singapore: NUS Press, 2011), 12.
[14] King, *Reading Bangkok,* 12, 33.
[15] King, *Reading Bangkok,* 123.
[16] Richard O'Connor, "Place, Power and Discourse in the Thai Image of Bangkok," in *The Society of Siam: Selected Articles on the Siam Society's Centenary*, ed. Chris Baker (Bangkok: The Siam Society, 2004), 382.
[17] O'Connor, "Place, Power and Discourse in the Thai Image of Bangkok," 383.
[18] O'Connor, "Place, Power and Discourse in the Thai Image of Bangkok," 383.
[19] O'Connor, "Place, Power and Discourse in the Thai Image of Bangkok," 383.

for most of the time it is destined to exist as a memory of an event that has already passed. If we apply the same logic to the city then we begin to see it not as an objectified, three-dimensional, knowable static entity that can be measured and contained in a cartographic form but rather as a never-ending repository of momentary performances and the memories of those performances that combine into a living image. This may seem a rather abstract methodology to discuss urban studies, but in the case of the highly animistic Thai culture it is not as far-fetched as one may think.

The chaotic, paradoxical and rather human-like character of Bangkok seems to be the common ground for anyone who has ever grappled with the task of describing the Thai City of Angels. In *Bangkok by Design*, John Hoskin complains: "Bangkok is not a beautiful city. … it sprawls across the unrelieved monotony of an alluvial plain in a way that was never planned and little expected."[20] Maryvelma O'Neill seems more understanding in her cultural history of the city explaining that "Bangkok makes more sense as an attitude than as a form."[21] She quotes Alexander McDonald, the founder of the longest running English language Thai newspaper, *The Bangkok Post*, who famously compared the city's affair with modernity to "a woman trying on hats, [who] oftentimes giggled at their absurdity and discarded them, relaxing to more informal ways of life."[22] S.P. Somtow, a Thai writer and composer, argues that Bangkok "is the one place in the world where there is absolutely no distinction between illusion and reality"[23] but, at the same time, refuses to see it in terms of a paradox, since, as he explains: "In Thailand, two contradictory things don't just have to *seem* to be true at the same time; they can actually *be* true! So these things aren't paradoxes at all."[24] He concludes with a sensual promise that "Bangkok … is a lover who can fulfil your every fantasy, and makes that fantasy utterly real for the duration of it; it is the one place where you actually can buy love."[25]

Once we begin to see Bangkok in terms of "afterthoughts," or "unintended consequences of the way Thai society works,"[26] things

---

[20] John Hoskin, *Bangkok by Design: Architectural Diversity in the City of Angels* (Bangkok: Post Books, 1995), 12.
[21] Maryvelma O'Neill, *Bangkok: A Cultural and Literary History* (Oxford: Signal Books, 2008), 3.
[22] Alexander McDonald quoted in O'Neill, *Bangkok: A Cultural and Literary History*, 33.
[23] S. P. Somtow, *The Other City of Angels* (Bangkok: Diplodocus Press, 2007), 91.
[24] Somtow, *The Other City of Angels,* 169.
[25] Somtow, *The Other City of Angels,* 91.
[26] O'Connor, "Place, Power and Discourse in the Thai Image of Bangkok," 400.

become even more complex, for the Thai society is a society of screens, regulated by what Peter Jackson calls "the Thai regime of images."[27] King devoted much of his analysis of Bangkok to the discussion of screens, which can never be questioned or peeled off but need to be seen through to grasp the multi-layering of Thai culture and its product, the city of Bangkok. Some screens are easy to see – you will find them in the form of glitzy advertising billboards, election posters, royal family portraits, swarms of unregulated private notice boards, or even the snake pit of overhanging cables triumphantly draped across every façade. Erected in the thousands, these kinds of screens frequently hide from view some more unsavoury sights. One well publicized example of such a screen was the wall constructed to hide the squatter slum settlement of Duang Phitak from the view of the participants of the World Bank/International Money Fund Conference that took place in the Queen Sirikit Conference Centre in 1991.[28] Ironically also, the mere existence of such screens provides the city inhabitants with convenient pockets of space, which in a short time fill up with all sorts of debris, magnifying our perception of Bangkok as the "face city," perpetuating the illusion of cleanliness by removing its dirt from view.

Bangkok screens, King argues, have also a functional dimension,[29] best characterised by the city's lenient attitude to informal economic activity (defined as non-formal production unregulated by law and not taxed, but also production of illegal goods and provision of illegal services), or indeed its dependence on it. According to the UN-affiliated International Labour Organization (ILO), in 2006, the informal economy in Thailand accounted for approximately 50% of the population and an estimated 45% of GDP.[30] The actual numbers will likely never be known. In 2012, the National Health Commission Office of Thailand estimated there were 4 million migrant workers in the kingdom, most of whom were

---

[27] Peter Jackson, "The Performative State: Semi-coloniality and the Tyranny of Images in Modern Thailand," *Sojourn* 19.2, (2004), <http://www.jstor.org/stable/41308172>, 219-253,

[28] Marc Askew, *Bangkok: Place, Practice and Representation* (London & NY: Routledge, 2002), 139.

[29] King, *Reading Bangkok*, 87.

[30] Christine Evans-Klock, "Foreword" in *Extending Labour Law to All Workers: Promoting Decent Work in the Informal Economy in Cambodia, Thailand and Mongolia*, ed. David Tajgman (Bangkok: International Labour Office, 2006), <http://www.ilo.org/asia/whatwedo/publications/WCMS_BK_PB_129_EN/lang--en/index.htm>, 5.

undocumented.[31] But perhaps the most pervasive kind of screens are the comforting and motivating symbolic superimpositions of the Thai triple ideal of the Nation, King and Religion, obscuring the fact that

> The peculiar form of modernity that manifests itself in Thailand is … built on epistemology that permits no (public) thought that would disturb the surface calm of the state …, nor any that would infringe the privacy of actions that might coincidentally [wreak] havoc on the skyline of the city, clog its roads or pollute its waterways.[32]

Any discussion of the rigid divide between the publicly unspeakable and privately obvious, which characterizes the modern surface-oriented Thai political system and a great deal of the country's cultural logic,[33] can be related to the notion of "face" or "masking" in which a performance of traditional cultural patters is required "to maintain the appearances of ideal cultural order."[34] Interestingly, both Rosalind Morris and Peter Jackson agree that one viable result of Thai modernization is a resurgence of magic, which in its post-1990s market-driven, technology-infused, politicized format has become a defining force in Thai urbanism, in continuous dialogue with the rationalized social order.[35]

Urban practices, like a great deal of Thai social life, are largely determined by popular religion, referred to by Pattana Kittiarsa as "a 'third space' where conventional Theravada Buddhism, state and *Sangha* authorities, multi-original religious beliefs and practices and the drive for material success in the capitalist market all come to coexist and produce a hybrid moment of religious change."[36] The spiritual climate of modern Bangkok is dominated by popular urban spirit-medium cults. Magic and mediums permeate private and public lives of contemporary Bangkokians. Astrologers, fortune-tellers, spirit mediums and Chinese geomancers are in themselves sought-after commodities. They solve heartbreaks, act as

---

[31] "The Number of Migrant Workers in Thailand Reaches 4 Million," (National Health Commission Office of Thailand, 28.11.2012), <http://en.nationalheal.or.th /node/294>.

[32] King, *Reading Bangkok*, 37.

[33] Jackson, "The Performative State: Semi-coloniality and the Tyranny of Images in Modern Thailand," 220.

[34] Rosalind Morris quoted in Jackson, "The Performative State: Semi-coloniality and the Tyranny of Images in Modern Thailand," 221.

[35] Jackson, "The Performative State: Semi-coloniality and the Tyranny of Images in Modern Thailand," 227-228.

[36] Pattana Kittiarsa, "Beyond Syncretism: Hybridization of Popular Religion in Contemporary Thailand," *Journal of Southeast Asian Studies* 36.3 (2005), 468.

career advisers, determine dates for auspicious events, and occasionally come up with lucky lottery numbers. There is nothing, it appears, they cannot handle – from matters of national security to such seemingly trifle issues as magically altering the colour of one's car.[37]

Behind Bangkok's spiritual landscape lies the Buddhist renunciation of illusion and the fact that, as King observes, "In a Thai episteme, there is always the problem of 'the real.'"[38] But in a sudden twist of an argument the Thais arrive at the conclusion which proclaims that "The world is ephemeral; like the cities we build and the wealth we accumulate, we will pass away. The world of the spirits, however, is forever."[39] And in pursuit of this spiritual infiniteness, they engage in magical rituals of temporality, nowhere as difficult to ignore as in the heart of a 10-million metropolis.

City Magic

Pile argues that "magic is not just *in the city*, but also *of the city*"[40] observing that magical beliefs can not only be found but are also created in urban settings. Bangkok's magic begins with the myth of its foundation, when according to a local legend, four commoners were buried alive under the city pillar (*lak muang*) as a human sacrifice, selected for the auspicious meaning of their names which combined into a prosperity charm. When read together, the names of the unfortunate villagers – Yuu, Yong, Man, Kong – aligned into a saying that roughly translated "this city shall stay strong for eternity." Marlane Guelden accounts for a similar story that replaces (un)luckily named commoners with soldiers, ready to defend the city in life and in death[41] and mentions other accounts claiming that further three victims were buried alive under Bangkok city gates in the early 1800s.[42]

While the city pillar may be simply understood in terms of the foundation point of the city, in the Thai context it is also the residence of

---

[37] One of the relatively recent additions to the Thai fortune-tellers' repertoire is the distribution of car stickers that magically "change" the colour of your car to a more "lucky" one. The stickers, which proclaim that "This car is blue, red, white, etc." are meant to be performative utterances ensuring spiritual protection and saving money on a paint job.

[38] King, *Reading Bangkok*, 114.

[39] King, *Reading Bangkok*, 114.

[40] Pile, *Real Cities: Modernity, Space and the Phantasmagorias of City Life*, 72.

[41] Marlane Guelden, *Thailand: Into the Spirit World* (Bangkok: Asia Books, 1995), 90.

[42] Guelden, *Thailand: Into the Spirit World*, 91.

the tutelary deity of the land. The Bangkok city pillar was erected by King Rama I in 1782. It was made from a local *chaiyapruek*[43] tree, which was said to contain a spirit of its own, and its spire entombed a gold plate inscribed with astrological predictions for the city. In 1995, Guelden reported that the shrine collected around 500 USD per day in offerings from a constant stream of visitors who engaged in a number of activities: from releasing caged birds, or draping gauze around pillar-like structures, to adding oil to continuously burning lamps, and sprinkling holy water over their heads.[44]

The pillar spirit may be the most important guardian of the city; it is, however, far from the only one. Although, as King points out, mechanisms of zoning and development control propagated in Western-style urban planning are incompatible with a Thai understanding of space,[45] this does not seem to be the case when the spiritual landscape is concerned. Spirit houses (*san phra poom*) of all sorts and sizes abound in the city, keeping guard over their clearly delineated allotments. They can be found hidden from view in laybys, *khlong*[46]-side communities, or private gardens, but can also take a prominent position in front of major shopping malls, luxury hotels, or government offices; they are erected near hospitals, factories, schools and universities, even on temple grounds, though officially worshipping guardian spirits has little to do with Buddhist doctrine. While popular religion may have been at times dismissed by the Thai authorities and Buddhist religious leaders as mere superstition, unacceptable in the rational dialogue on modernization, the pervasiveness of urban spirit-medium cults, a single nodule in the complex network of modern animistic practices, seems to prove otherwise. Kittiarsa cites the study of Ekkachai Sanitsuda estimating that by the 1990s there were "more than 100,000 mediums in Thailand whose services generated over 20 billion *baht* per year."[47] If nothing else, this is a significant boost to Bangkok's informal economy and deserves further consideration.

Much has been written on the materialism of Thai popular religion and its relationship to urbanism and the financial practices of Thai capitalism. Peter A. Jackson explored the Thai religions of prosperity, or "popular movements that emphasize the acquisition of wealth as much as

43 *Cassia javanica*, commonly known as Pink Shower tree.

44 Guelden, *Thailand: Into the Spirit World*, 91.

45 King, *Reading Bangkok*, 127.

46 A *khlong* is a small canal. Many of Bangkok's roads used to be canals in the past.

47 Ekkachai Sanitsuda quoted in Kittiarsa, "Beyond Syncretism," 471.

salvation"[48] that have become prominent since the 1990s. Alan Klima wrote about the association of mediumistic practices with post-1997 underground lotteries, gambling and personal moneylending.[49] Shrines never close on lottery nights and predictions of the winning numbers can be devised in a variety of ways, through pasting thin gold leaf squares onto the effigy of the deity, shaking a can of Chinese *kau cim* fortune sticks, rubbing camphor oil into the bark of an auspicious tree, or playing with fireworks, amongst others. As Thomas Fuller reports in *New York Times*, inspiration for choosing a specific number may come from the most unusual of sources, like spotting an animal born with an extra appendage, or betting on a celebrity's hotel room number.[50] Indeed, in March 2013, a resident of Buri Ram, proudly declared that a kitten corpse he found in a local sausage was a dwelling of not one but nine goddesses, all of whom were well-versed in lottery matters.[51] But most Thai people seem to flock to auspicious places that are said to be inhabited by knowledgeable spirits, often powerful ghosts who died in tragic circumstances. Such shrines can be found all over Bangkok. Fuller quotes three, which, according to his sources, all belong to the Thai "top ten": the shrine to victims of road accidents locally known as "The Tree of 100 Corpses," the shrine of Thailand's most beloved ghost, Mae Nak Phra Khanong, and, apparently, the fourth floor of a police station in Bangkok's Chinatown.[52]

While the lottery, whether government-run or underground, can be seen as contributing to "grey area" economics, things are not so different for legitimate commerce. One of the most prominent shrines, located in front of the luxurious Grand Hyatt Erawan hotel and shopping mall at the busy Ratchaprasong intersection in the heart of commercial Bangkok, shares the name with the complex. Dedicated to the Hindu god, Brahma, the shrine (also known as *Thao Maha Phrom* in Thai) was erected in 1957 after a consultation with a Brahmanic astrologer to put an end to a series of

[48] Peter A. Jackson, "Royal Spirits, Chinese Gods, and Magic Monks: Thailand's Boom-Time Religions of Prosperity," *South East Asia Research* 7.3 (1999), 245.
[49] Alan Klima, "Spirits of 'Dark Finance' in Thailand: A Local Hazard for the International Money Fund," *Cultural Dynamics* 18.33 (2006), <http://cdy.sagepub.com/content/18/1/33.full.pdf+html>, 33-60.
[50] Thomas Fuller, "Plane Crash? Murders? Time to Play Thai Lottery," *New York Times* (January 2, 2013), <http://www.nytimes.com/2013/01/03/world/asia/deadly-car-crash-time-to-play-the-thai-lottery.html?_r=0>.
[51] "Owner Says Nine Goddesses Dwell in Divine Kitten Sausage," *Coconuts Bangkok* (March 19, 2013), <http://www.coconutsbangkok.com/news/owner-says-nine-goddesses-dwell-in-divine-kitten-sausage/>.
[52] Fuller, "Plane Crash? Murders? Time to Play Thai Lottery."

misfortunes that plagued the construction of the hotel. Erawan Shrine is not the only spiritually-informed fail-safe system in the area. On the opposite side of the road, in front of Central World Plaza stands another pronounced shrine to the Hindu Holy Trinity (Vishnu, Siva and Brahma), while round the corner, near the Intercontinental Hotel on Witthayu Rd., you will find a tree-shrine to Goddess Tuptim, hidden amongst a multitude of wooden phalluses (*lingams*). Still, Erawan Shrine seems to dwarf its neighbours when it comes to popularity, commonly believed to be so auspicious for business that fees for dancers entertaining the deity can now be deducted from taxes. Its Hindu god is rumoured to have a taste for nude dancing, which occasionally is said to lead to devotees flashing their breasts at the statue at night, or donating pornographic videos instead.[53] But for most times of the day, spiritual business is taken very seriously.

On 21 March 2006, Erewan Shrine was smashed with the hammer by a 27-year-old man, Thanakorn Pakdeepol. The culprit was apprehended on the spot and beaten to death by angry bystanders who were later charged with homicide and almost immediately released,[54] while the locals generally agreed that the vandal got what he deserved.[55] Although Pakdeepol was reported to be mentally ill, and his attack on the shrine could well be seen as accidental, it was also publicly interpreted as an omen relating to the crisis surrounding the then Prime Minister, Thaksin Shinawatra, and in effect foretelling the military coup of 19 September 2006. As Charles Keyes observes: "Both private and public discourses about spirits, omens, and horoscopes contribute to the understandings Thai have about personal affairs and public political crises."[56] He concludes that although the destruction of the shrine did not directly cause Thaksin's fall from power, it nevertheless contributed to it, by seriously compromising the PM's authority.[57]

---

[53] Guelden, *Thailand: Into the Spirit World*, 10.
[54] Monthathip Poo "Man Beaten to Death after Desecrating the Erawan Shrine, *The Nation* (March 22, 2006), <http://www.nationmultimedia.com/2006/03/22/headlines/headlines_20003314.php>.
[55] King, *Reading Bangkok*, 114.
[56] Charles Keyes, "The Destruction of a Shrine to Brahma in Bangkok and the Fall of Thaksin Shinawatra: The Occult and the Thai Coup in Thailand of September 2006," *ARI Working Paper No. 80* (2006), <http://www.ari.nus.edu.sg/showfile.asp?pubid=638&type=2>, 27.
[57] Keyes, "The Destruction of a Shrine to Brahma in Bangkok and the Fall of Thaksin Shinawatra: The Occult and the Thai Coup in Thailand of September 2006," 27.

Pasuk Phongpaichit and Chris Baker take up the story of magic and Thai politics exactly where Keyes has left it. In their paper delivered at the Siam Society in December 2008, the authors pile up evidence of astrological and shamanistic practices of major Thai political figures. Prominent military players are identified as reincarnations of famous dead generals[58]; others have to rely of their wives inseparability from a pink toy elephant.[59] Astrologers and fortune-tellers are consulted almost on a daily basis and certainly before any important events; some politicians are even reported to be resorting to Khmer black magic and are willing to vandalise an ancient temple in the process.[60] Phongpaichit and Baker point out the long association of astrology with political power in Thailand, its authority being derived from old royal tradition. They also observe that astrology is passive and therefore has its limitations, and as such is often abandoned by some more pro-active politicians for the sake of other techniques that may allow them to "improve" their fortune instead.[61] *Saiyasat* (the general term for supernaturalism) brings together the old belief in Vedic incantations and magical objects, the belief in achieving mastery through cultivating supernatural abilities, and the belief in the interference of the spirits of the dead in the world of the living.[62] Magical tattoos, amulets, spiritual ceremonies, wax or clay magic dolls or "missiles" planted within the living flesh of one's enemy, are just part of the supernatural repertoire available to ordinary, as well as less ordinary citizens of Bangkok.

Both past and current Thai political clashes have been punctuated by popular religious and magical rituals playing to their respective mass audiences. During the 2008 anti-government protests of the People's Alliance for Democracy (PAD), the protesters "blocked" the spiritual power of statues around the Government's House by ceremonially "closing" their eyes, or even more drastically, by surrounding them strategically with used tampons and sanitary napkins, as menstrual blood is generally believed to render magic useless.[63] In an even more theatrical gesture, in 2010 the United Front for Democracy against Dictatorship (UDD) performed a Brahmin blood ritual which involved pouring litres of human blood in front of the Government House and the Prime Minister's

---

[58] Pasuk Phongpaichit and Chris Baker, "The Spirits, the Stars and Thai Politics," an unpublished paper delivered at Siam Society Conference (December 2, 2008), <http://pioneer.netserv.chula.ac.th/~ppasuk/spiritsstarspolitics.pdf>, 1.
[59] Phongpaichit and Baker, "The Spirits, the Stars and Thai Politics," 2.
[60] Phongpaichit and Baker, "The Spirits, the Stars and Thai Politics," 4.
[61] Phongpaichit and Baker, "The Spirits, the Stars and Thai Politics," 8.
[62] Phongpaichit and Baker, "The Spirits, the Stars and Thai Politics," 9.
[63] Phongpaichit and Baker, "The Spirits, the Stars and Thai Politics," 6.

house.[64] On 20 March 2010, Taliesin Verity reported for *Pattaya Daily News:* "The Brahmin who had been employed to twice curse the Democrats was later denounced by the Royal Brahminical leadership as not being in keeping with Brahmin traditions and even questioning his right to call himself a Brahmin; which he countered by saying that his father had been a Royal Brahmin priest."[65]

"At the core of dreams and magic are (unfulfilled) wishes,"[66] wrote Pile. This, in itself orientates the magical act towards the future. Its practices are at the same time "emotional, expressive, gestural, dramatic, impulsive and performative."[67] They are meant to satisfy individual wishes. City magic has its links with the three big forces shaping human societies: Law, Politics and Economics. Whether harnessing cosmological energy, using magical influence in the economy, or directly manipulating space to affect conditions for work and habitation, city magic has the power to transform the energy, spirit and physical shape of the city. The construction of all the major structures in Bangkok has always been likely to begin from consultations with astrologers and geomancers. Indeed, the failure of some projects is often attributed to their contradictory or inauspicious *feng shui*. This was certainly the case with many unfinished buildings that fell victim to the 1997 crisis, and whose empty ruins still punctuate the city's skyline. Occasionally, the advantageous positioning of buildings may be abandoned to accommodate the architect's folly, like in the case of the famous Bangkok landmark, the Chang Building on Phahonyothin Rd. in the shape of a charging elephant that seems to constantly threaten its surroundings.[68] They can also occasionally be counteracted by a stronger magic: Central World Plaza burnt down during the 2010 anti-government demonstrations despite its elaborate *feng shui* protection.

Bangkok magic comes to life through a never-ending series of ritualistic performances fuelled by hopes of improving the future. Bangkok

---

[64] "PM's House Splattered with Blood," *Asia Views* (December 21, 2010), <http://www.asiaviews.org/index.php?option=com_content&view=article&id=27513:storyalias40> and "Red Shirts Take Blood Protests to PM," *Al Jazeera* (March 17, 2010), <http://www.aljazeera.com/news/asia-pacific/2010/03/201031733210935382.html>.

[65] Taliesin Verity, "The Underlying Magic behind the Thai Battle for Power," *Pattaya Daily News* (March 20, 2010), <http://www.pattayadailynews.com/en/2010/03/20/the-underlying-magic-behind-the-thai-battle-for-power/>.

[66] Pile, *Real Cities: Modernity, Space and the Phantasmagorias of City Life*, 59.

[67] Pile, *Real Cities: Modernity, Space and the Phantasmagorias of City Life,* 63.

[68] King, *Reading Bangkok*, 188-189.

may be a dynamic 10-million commercial metropolis but it is also a place filled with holy trees and termite mounds, shrines, altars and spirit houses of all shapes and sizes, adorned with clay dolls, wooden phalluses, and effigies of spirits and wise men. It is also a space realized through magic and spirit worship, performed through a million acts of feeding the spirits, rubbing the trees, praying to snakes, raising ghostly babies, consulting witch-doctors and fortune-tellers, performing exorcisms, inscribing one's body with magical tattoos, collecting protective amulets, casting love-spells, or throwing curses. Each promise has the potential to change the city; each results in the creation of a temporary vision of the city's future. Each magical enunciation contributes to making the city alive.

Walking with Ghosts

Buddhist cultures are not supposed to dwell on the past. Thai historical narrative does not admit the discussion of events that hint at national disharmony. The excuse of maintaining social unity seen as a behavioural imperative has led to a widespread practice of self-censorship and avoidance of mentioning any potentially contestable topics. And yet Bangkok is alive with ghosts, walking alongside the living, seemingly unaware of their demise. Ghosts "betray the significance of time and memory in the production of urban spaces," wrote Pile.[69] Their random materialization "destabilize[s] the flow of time of a place … They change a place's relationship to the passage of time – as ghosts seem to pop out of nowhere, but are somehow always there."[70] But ghosts can also leave their mark on the city landscape. They can be responsible for an entire district falling into disuse, while other territories thrive. They can account for an area's both good and ill repute.

Take New Petchaburi Rd. in the heart of Bangkok as an example, a prime location that seems to resist development, scaring the passers-by with the hollow stare of its empty townhouses. In 1990, this was the site of perhaps the most devastating road accident in the country's history, involving the explosion of a truck with two 20,000-litre liquid petroleum gas tanks, that left 59 people dead, hundreds injured, destroyed 43 cars and 4 motorbikes, and ravaged 21 city blocks. The fires were reported burning for more than 24 hours.[71] In her account of the event, Guelden mentions a

---

[69] Pile, *Real Cities: Modernity, Space and the Phantasmagorias of City Life*, 131.
[70] Pile, *Real Cities: Modernity, Space and the Phantasmagorias of City Life,* 131.
[71] "20th Anniversary of Devastating New Petchaburi Road Gas Explosion," *Pattaya Daily News* (September 28, 2010), <http://www.pattayadailynews.com/en/2010/09/28/20th-anniversary-of-devastating-new-petchaburi-road-gas-explosion/>.

significantly higher death toll of 89, with 51 cars and 61 buildings destroyed.[72] She also notes that for seven days after the accident drivers were alerted to avoid the area as it is common knowledge that ghosts of road accident victims tend to cause more accidents. The city responded with organising a Buddhist ceremony on the accident site. While the monks chanted prayers for the departed, a rational explanation for the accident was encouraged, and the living were advised to help the families of the victims rather than waste time and money on superstitions. In a blatant disregard of the official sentiments, on the opposite side of the road, offerings of cooked food, clothes and paper money were being steadily delivered to appease the hungry spirits of the violently dead (*phii tai hong*), monstrous spiritual entities doomed to a ghostly afterlife by their untimely death.[73]

The angry ghosts of the revellers who burnt to death celebrating New Year's Eve in a trendy Bangkok club, *Santika*, on 1st Jan 2009 have been sighted in the fashionable district of Thong Lor, but this seemed to have little effect on the local club-goers, significantly more interested in worshipping spirits of the alcoholic kind. Legitimized by its aura of celebrity, Thong Lor seems quite resistant to supernatural menace. The same cannot be said about the area around Wat Phai Ngern, a small temple situated in the Bang Kho Laem district in central Bangkok, put in the spotlight in 2010 due to a grisly discovery of 2,002 aborted human foetuses in its storage lockers. Since abortion is illegal in Thailand, disposing of its evidence can be a lucrative business, although the penalty it carries can be seen as discouraging by some (the undertaker convicted for concealing the foetuses was originally sentenced to 667 years in prison, the term finally shortened to 20 years).[74] In the days following the discovery, Bangkok residents flocked to the temple grounds, some appalled by the event, others hoping there may be some luck born out of this misfortune. Though abortion may be an unmentionable subject in Thailand, cultivating baby spirits (*kuman tong*), originally obtained through a rather gruesome ritual which involved grilling human foetuses and entombing them in a jar, is still an on-going practice. Even though the monks insisted that the temple foetuses were not going to be saved for that

---

[72] Guelden, *Thailand: Into the Spirit World*, 49.

[73] Guelden, *Thailand: Into the Spirit World,* 55-56.

[74] "Bangkok, Undertaker Jailed for 20 Years for Concealing 2002 Foetuses," *AsiaNews.It* (January 20, 2011), <http://www.asianews.it/news-en/Bangkok,-undertaker-jailed-for-20-years-for-concealing-2002-foetuses:-20625.html>.

purpose,[75] the association of baby spirits with good fortune did wonders for the sales of lottery tickets.

Just as every piece of land in Bangkok is commonly believed to have its spiritual guardians, every one of them is said to have occasionally been haunted by ghosts. At ease with the modern city life, Bangkok ghosts have been spotted on buses, or riding the *Skytrain*, they have been known to linger around car parks, in shops and office elevators, even Suvarnabhumi Airport, it has been reported, is not safe from spectral intrusion.[76] The violently dead - victims of accidents, suicides and murders – account for a large majority of the sightings, but they are not the only ones. Ghosts, argues Michael Mayerfeld Bell, "that is, *the sense of the presence of those who are not physically there* – are a ubiquitous aspect of the phenomenology of place. … we constitute a place in large measure by the ghosts we sense inhabit and possess it.[77] Bell defines a "place" as a space enhanced with a social meaning and sees ghosts as instrumental for this process.[78] The belief that ghosts haunt certain spaces or objects helps us experience them socially. "Through ghosts, we re-encounter the aura of social life in the aura of place."[79]

Nowhere is this more strongly felt than in the Bangkok district of Phra Khanong, associated with the most popular Thai ghost, Mae Nak. According to one version of the story, Mae Nak died in childbirth while her husband was at war. Upon the husband's return, the ghost of Nak resumed her wifely duties, at the same time terrorizing the villagers to keep them from revealing the truth about her true nature to her unsuspecting partner. The spectral wife was finally subdued by a famous monk who turned part of her skull into a potent talisman and entombed her in a clay pot, but it seems that her spirit cannot truly be put to rest, being resurrected in a multiplicity of narratives. In the late nineteenth century, Prince Damrong, a reputable Thai historian, conducted a poll at the gates of the Grand Palace, which revealed that Bangkokians knew more about

---

[75] "Phai Ngern Saga: No Plan to Turn Carcasses of Dead Babies into Guman Thong," *Thai-Asean News Network* (January 13, 2012), <http://www.tannetwork.tv/tan/ViewData.aspx?DataID=1051561>.

[76] Martha Jette, "Blue Ghost Haunted Bangkok Airport," *Gather*, (January 28, 2007), <http://www.gather.com/viewArticle.action?articleId=281474976896773>.

[77] Michael Mayerfeld, Bell, "The Ghosts of Place," *Theory and Society* 26.6 (1997), <http://links.jstor.org/sici?sici=0304421%28199712%2926%3A6%3C813%3ATGOP%3E2.0.CO%3B2-E>, 813.

[78] Bell, "The Ghosts of Place," 820.

[79] Bell, "The Ghosts of Place," 821.

the life of Mae Nak than about King Rama I.[80] Her tale of woe, told and retold by generations of Thais, has already re-emerged as a poem, a novelette, a play, a radio drama, a musical, an opera, a cartoon, and over 25 movies, including at least one in 3D. Although accounts of the story differ and there is no factual evidence that a woman called Nak even existed, not to mention returned as a ghost, her effigy, venerated in a shrine as a local deity, continues to receive daily doses of offerings from crowds of devotees asking her for advice on love matters, protection from military conscription, or simply hoping to get a tip on lucky lottery numbers.

The accounts like that of a Thai historian, Anake Nawikamune, who claimed that the story of Mae Nak originated from a fake haunting staged by the sons of a woman known as Amdaeng Nak, who by appearing as a ghost of their mother, were trying to prevent their father from remarrying and thus forcing them to share their inheritance,[81] are chiefly ignored by the believers, whose faith in the ghostly presence of Mae Nak is unchallenged. In a shrine, adjacent to the Buddhist temple of Wat Mahabute, a funeral for the ghost is performed repeatedly once a month.[82] Piles of the offerings for Mae Nak and her baby – plastic toys, dresses, cosmetics – seem to cover every inch of the space while the ghost itself (inhabiting its gold-leafed effigy) sits transfixed watching television soap operas all day long. McDaniel sees in Mae Nak's shrine an example of what Pierre Nora refers to as sites of memory, "the ultimate embodiments of a memorial consciousness" that "mark the rituals of a society without ritual."[83] For Nora the sites of memory are "moments of history torn away from the movement of history, then returned; no longer quite life, not yet death, like shells on the shore when the sea of living memory has receded."[84] McDaniel's discussion of the people, texts, actions and objects associated with Mae Nak inscribes those particular sites of memory within the local consciousness of Bangkok.

---

80 O'Neill, *Bangkok: A Cultural and Literary History,* 13.

81 "The Legend of Mae Nak Prakanong," *Sarakadee Magazine* (September 1999), <http://www.sarakadee.com/feature/1999/09/nang-nak.htm>.

82 Justin Thomas McDaniel, *The Lovelorn Ghost & The Magical Monk* (NY: Columbia UP, 2011), 3.

83 Pierre Nora, "Between Memory and History: *Les Lieux de Mémoire*," *Representations* 26 (1989), 12.

84 Nora, "Between Memory and History: *Les Lieux de Mémoire*," 12.

Maurice Halbwachs insisted that "every collective memory unfolds within a spatial framework"[85] because "[m]ost groups … engrave their form in some way upon the soil and retrieve their collective remembrances within the spatial framework thus defined."[86] Echoes of individual and collective traumas, which in themselves can be seen as group-forming, resonate through Bangkok spaces giving up their ghosts. High-profile cases involving large numbers of casualties, or specifically graphic details are believed to be more likely to result in hauntings, although victims of political violence are frequently denied that privilege in Thailand. Just like memorials and monuments, ghost-inhabited spaces are the sites of memory. Victims of the clashes with the government forces are seen as contesting history and as such are expected to be forgotten. In the twentieth century, Thailand saw between 18-20 military coups (the numbers vary depending on historical interpretation), often preceded or followed by prolonged periods of rioting and resulting in a significant death toll. Some of those events, like for instance the massacre of the students at Thammasat University in October 1976, "[do] not fit the normative ideology in Thailand for the memorable past" and as such they "[resist] comprehensibility and remembering."[87] They are not only denied official recognition, but it seems are even incapable of producing a ghost.

Thammasat University is said to have a haunted red-door elevator where on some occasions you can still see the ghosts of students shot by the militant troops in 1976. But the very few existing historical records of the day tell a different story and the local memory of the event has been very successfully supressed. While the red-door elevator has its place in the spectral landscape of the city, it seems to exist only here and now, its historical significance having been pushed into silence. This is not entirely unexpected. "However we locate them temporally, the ghosts of place are always *presences* and as such appear to us as spirits of temporal transcendence, of connection between past and future."[88] Even when the past gets silenced, its ghosts can reappear in the present. Their appearance is never a permanent or even a particularly reliable phenomenon, but through their presence we experience transient moments of temporal

---

[85] Maurice Halbwachs, *The Collective Memory*, trans. Francis J. Ditter, Jr. and Vida Yazdi Ditter (New York: Harper & Row, 1980), 140.

[86] Halbwachs, *The Collective Memory*, 156.

[87] Thongchai Winichakul, "Remembering/Silencing the Traumatic Past: The Ambivalent Memories of the October 1976 Massacre in Bangkok," in *Cultural Crisis and Social Memory: Modernity and Identity in Thailand and Laos,* ed. Shigeharu Tanabe and Charles F. Keyes (London & NY: Routledge, 2002), 246.

[88] Bell, "The Ghosts of Place," 816.

collision and the haunted space exists simultaneously in the past and the present. Every urban space is alive with thousands of ghosts willed to life by the passers-by in an act of its socialization. Seen as symbolic representations of individual or collective memory (often repressed memory), ghosts reappear to help us come to terms with past traumas, escape into the nostalgic fantasy of our childhood, or simply to give meaning to a place. In Bangkok, ghosts just happen to be a little bit more ghostly than that.

Thai popular religion, being infused with spiritualism and animistic beliefs does not question the existence of ghosts. Indeed, in a land filled with real ghosts, it is their symbolism that frequently meets with resistance. A Thai saying warns: "You may not believe, but never offend the spirits." But whether ghosts do walk the streets of Bangkok, or they are simply being conjured up by Bangkokians humanizing their neighbourhood, their by-product is the second city trapped in a time rift which reveals itself through each and every spiritual interaction. This second Bangkok, a collection of ephemeral moments, comes alive with each active act of ritualistic magical performance and each passive encounter with its spectral past, which to the Thais are two sides of the same coin. To Bangkokians, this temporal city is infinitely more important than its image captured on a map, defined in official documents, or rendered through any other legitimate forms of representation. But we already know that. After all, the world is just illusion and "[t]he world of the spirits … is forever."[89]

References

Author Unknown, 1999, "The Legend of Mae Nak Prakanong," *Sarakadee Magazine*, <http://www.sarakadee.com/feature/1999/09/nang-nak.htm>.

Author Unknown, 2006, "Man Beaten to Death after Desecrating the Erawan Shrine." *The Nation*, <http://www.nationmultimedia.com/2006/03/22/headlines/headlines_20003314.php>.

Author Unknown, 2010, "20th Anniversary of Devastating New Petchaburi Road Gas Explosion." *Pattaya Daily News*, <http://www.pattayadailynews.com/en/2010/09/28/20th-anniversary-of-devastating-new-petchaburi-road-gas-explosion/>.

[89] King, *Reading Bangkok*, 114.

Author Unknown, 2010, "PM's House Splattered with Blood," *Asia Views*, <http://www.asiaviews.org/index.php?option=com_content&view=article&id=27513:storyalias40>.

Author Unknown, 2010, "Red Shirts Take Blood Protests to PM," *Al Jazeera*, <http://www.aljazeera.com/news/asia-pacific/2010/03/201031733210935382.html>.

Author Unknown, 2011, "Bangkok, Undertaker Jailed for 20 Years for Concealing 2002 Foetuses." *AsiaNews.It*, <http://www.asianews.it/news-en/Bangkok,-undertaker-jailed-for-20-years-for-concealing-2002-foetuses:-20625.html>.

Author Unknown, 2012, "Phai Ngern Saga: No Plan to Turn Carcasses of Dead Babies into Guman Thong," *Thai-Asean News Network*, <http://www.tannetwork.tv/tan/ViewData.aspx?DataID=1051561>.

Author Unknown, 2012, "The Number of Migrant Workers in Thailand Reaches 4 Million," National Health Commission Office of Thailand, <http://en.nationalhealth.or.th/node/294>.

Author Unknown, 2013, "Owner Says Nine Goddesses Dwell in Divine Kitten Sausage," *Coconuts Bangkok*, <http://www.coconutsbangkok.com/news/owner-says-nine-goddesses-dwell-in-divine-kitten-sausage/>.

Askew, M., 2002, *Bangkok: Place, Practice and Representation*, London & NY: Routledge.

Bell, M. M., 1997, "The Ghosts of Place," *Theory and Society* 26.6, <http://links.jstor.org/sici?sici=0304421%28199712%2926%3A6%3C813%3ATGOP%3E2.0.CO%3B2-E>, 813-836.

Certeau, M. de., 1997, *The Practice of Everyday Life*, Rendall, S., trans., Berkley: University of California Press.

Fuller, T., 2013, "Plane Crash? Murders? Time to Play Thai Lottery," *New York Times*, <http://www.nytimes.com/2013/01/03/world/asia/deadly-car-crash-time-to-play-the-thai-lottery.html?_r=0>.

Guelden, M., 1995, *Thailand: Into the Spirit World,* Bangkok: Asia Books.

Halbwachs, M., 1980, *The Collective Memory*, Ditter, F. J. Jr., and Ditter, V. Y., trans., New York: Harper & Row.

Hall, E. T., 2003, "Monochromic and Polychromic Time," in Samovar, L. A., and Porter, R. E., eds., *Intercultural Communication: A Reader*, Belmont CA: Wadsworth/Thomson Learning, 262-268.

Hoskin, J., 1995, *Bangkok by Design: Architectural Diversity in the City of Angels*, Bangkok: Post Books.

Jackson, P. A., 1999, "Royal Spirits, Chinese Gods, and Magic Monks: Thailand's Boom-Time Religions of Prosperity," *South East Asia Research* 7.3, 245-320.

Jackson, P., 2004, "The Performative State: Semi-coloniality and the Tyranny of Images in Modern Thailand," *Sojourn* 19.2, <http://www.jstor.org/stable/41308172>, 219-253.

Jette, M., 2007, "Blue Ghost Haunted Bangkok Airport," *Gather*, <http://www.gather.com/viewArticle.action?articleId=281474976896773>.

Keyes, Ch., 2006, "The Destruction of a Shrine to Brahma in Bangkok and the Fall of Thaksin Shinawatra: The Occult and the Thai Coup in Thailand of September 2006," *ARI Working Paper* No. 80, <http://www.ari.nus.edu.sg/showfile.asp?pubid=638&type=2>, 1-30.

King, R., 2011, *Reading Bangkok*, Singapore: NUS Press.

Kittiarsa, P., 2005, "Beyond Syncretism: Hybridization of Popular Religion in Contemporary Thailand," *Journal of Southeast Asian Studies* 36.3, 461-487.

Klima, A., 2006, "Spirits of 'Dark Finance' in Thailand: A Local Hazard for the International Money Fund," *Cultural Dynamics* 18.33, 33-60.

Makeham, P., 2005, "Performing the City," *Theatre Research International* 30.2, 150-160.

McDaniel, J. T., 2011, *The Lovelorn Ghost & The Magical Monk*, NY: Columbia UP.

Mumford, L., 2009, "What is a City," in LeGates, R.T., and Stout, F., eds., *The City Reader* 4th Edition, London & NY: Routledge. 85-89.

Nora, P., 1989, "Between Memory and History: *Les Lieux de Mémoire*," *Representations* 26, 7-25.

O'Connor, R., 2004, "Place, Power and Discourse in the Thai Image of Bangkok," in Baker, Ch., ed., *The Society of Siam: Selected Articles on the Siam Society's Centenary*, Bangkok: The Siam Society, 382-405.

O'Neill, M., *Bangkok: A Cultural and Literary History*, Oxford: Signal Books, 2008.

Park, R. E., 2004, "Suggestions for the Investigation of Human Behaviour in the Urban Environment," in Jenks, Ch., ed., *Urban Culture: Critical Concepts in History and Cultural Studies*, Vol. 1, London & NY: Routledge, 19-54.

Phongpaichit, P., and Baker, Ch., 2008, "The Spirits, the Stars and Thai Politics," An unpublished paper delivered at Siam Society, <http://pioneer.netserv.chula.ac.th/~ppasuk/spiritsstarspolitics.pdf.>.

Pile, S., 2005, *Real Cities: Modernity, Space and the Phantasmagorias of City Life*, London: Sage.

Somtow, S. P., 2007, *The Other City of Angels,* Bangkok: Diplodocus Press.

Tajgman, D., 2006, "Extending Labour Law to All Workers: Promoting Decent Work in the Informal Economy in Cambodia, Thailand and Mongolia," International Labour Organisation, <http://www.ilo.org/asia/whatwedo/publications/WCMS_BK_PB_129_EN/lang--en/index.htm>, 1-158.

Winichakul, T., 2002, "Remembering/Silencing the Traumatic Past: The Ambivalent Memories of the October 1976 Massacre in Bangkok," in Tanabe, Sh., and Keyes Ch. F., eds., *Cultural Crisis and Social Memory: Modernity and Identity in Thailand and Laos,* London & NY: Routledge, 243-283.

Verity, T., 2010, "The Underlying Magic behind the Thai Battle for Power," *Pattaya Daily News*, <http://www.pattayadailynews.com/en/2010/03/20/the-underlying-magic-behind-the-thai-battle-for-power/>.

READINGS

CHAPTER FOUR

REMEMBERING THE LOSS, CONSTRUCTING THE FUTURE: TIME AND MEMORY IN HELEN HUNT JACKSON'S *RAMONA*

KATARZYNA NOWAK-MCNEICE

> "Since the days when the fleet of Columbus sailed into the waters of the New World, America has been another name for opportunity, and the people of the United States have taken their tone from the incessant expansion which has not only been open but has even been forced upon them. He would be a rash prophet who should assert that the expansive character of American life has now entirely ceased. Movement has been its dominant fact, and, unless this training has no effect upon a people, the American energy will continually demand a wider field for its exercise. But never again will such gifts of free land offer themselves."[1]

The so called "Frontier thesis" Frederick Jackson Turner formulated in 1893 explains American character as well as American institutions in relation to the frontier, which, as Turner makes clear, is the single most important point of reference whose consideration is crucial for the understanding of American identity. In an important gesture, Turner turns away from Europe and looks at American history from within. It is neither Europe that provides a backdrop against which the drama of American history is played out, nor is it New England that is seen as the cradle of American identity. Turner claims that one must look to the frontier, this ever-progressing division, to find the uniqueness of the American nation.

[1] Frederick Jackson Turner, "The Significance of the Frontier in American History" in *The Frontier in American History* (New York: Henry Holt and Company, 1893), 38.

Naming and celebrating American characteristics such as resourcefulness and rugged individuality, Turner links their genealogy to the frontier, at the same time giving Americans an opportunity to see their history and national identity as exceptional: it is not based on memory and a look backwards, but the very project of identity is perceived as depending on forgetting the past and focusing on the future.

Such a rejection of the past is problematic, if only for the fact that it silences many agents of history. "The voices of race, class and gender were muted or absent in Turner's essay,"[2] Martin Ridge mournfully notes in an otherwise celebratory discussion of Turner's thesis, and such an accusation is but one of many that can be produced against Turner's obviously chauvinistic understanding of history. Undoubtedly, Turner's formulations shaped American understanding of the frontier and its place in both American history and imagination.

Turner's essay, though, was not entirely triumphant in tone: he underpinned the exposition of the frontier's significance for American identity with a rather bleak prospect for the future. Turner made it quite clear that the frontier had closed by the time he wrote the essay, and that the opportunity of an unhindered progress and a promise of unlimited expansion that the frontier meant for Americans would not repeat themselves. With the closing of the frontier, the era of unprecedented growth ended. Noble sums up Turner's pessimism: "Following close behind the westward expansion of freehold property came the westward expansion of an industrialism imported from Europe. Its property patterns, like those of commerce, were the corrupting ones of power and dependence."[3] The fear expressed is not only of following in the footsteps of Europe once again. It is the fear of America losing its uniqueness; a uniqueness that largely depends on America's turn to the future. The fear, in fact, may be seen as a consequence of this turn: the price that must be paid for the obliteration of memory.

Turner's thesis influenced generations of scholars discussing the American frontier: the causal relationship between the frontier and American character has been established and, despite criticism, never fully abandoned. What the frontier thesis also suggested was that at the core of American identity lies a loss: "never again," says Turner, "will such gifts

---

[2] Martin Ridge, "The Life of an Idea: Frederick Jackson Turner's Frontier Thesis," *Montana: the Magazine of Western History*, vol. 41 no. 1 (Winter 1991), 10.

[3] David W. Noble, *The End of American History: Democracy, Capitalism, and the Metaphor of Two Worlds in Anglo-American Historical Writing, 1880-1980* (Minneapolis: University of Minnesota Press, 1985), 22.

of free land offer themselves,"[4] intimating that it is in the act of mourning that American identity comes into being.

The link between mourning and identity was first established by Freud in his essay "Mourning and Melancholia" (1917) and then developed in *The Ego and the Id* (1923), with numerous critics following in his lead and commenting on the idea. Judith Butler builds on the Freudian understanding of identity when she mentions what she suspects is the most difficult of all losses, "the loss of loss itself: somewhere, sometime, something was lost, but no story can be told about it; no memory can retrieve it," yet paradoxically it can provide "the condition of a new political agency."[5] This she calls "a melancholic agency."[6] She elaborates on the social and political aspect of melancholy: "Loss becomes condition and necessity for a certain sense of community, where community does not overcome the loss, where community *cannot* overcome the loss without losing the very sense of itself as community."[7] Not only does the loss, then, become central to individual identity, but it also provides the basis for a communal sense of togetherness. Not overcoming the loss, which is formative of identity here, can be understood as memory itself: what is difficult to remember, yet what cannot be forgotten, generates and sustains identity.

In Helen Hunt Jackson's *Ramona,* a novel published in 1884, loss is at the core of the narrative. Set in California after it became an American state, it presents a romance between the title character, a half-Scottish, half-Indian girl, and a Native American named Alessandro, against the backdrop of a gradual dispossession of the Native American groups and their removal from their ancestral lands. Jackson knew the circumstances firsthand: she had visited Native American dwellings and interviewed the people. The situation so enraged the author that she considered it her moral obligation to campaign personally and through her writing to amend the injustice the Natives suffered at American hands. She did so till the last moments of her life; on her deathbed she wrote in a letter to the president,

---

[4] Jackson Turner, "The Significance of the Frontier in American History," 38.

[5] Judith Butler, "Afterword: After Loss, What Then?" in *Loss: The Politics of Mourning,* eds. David L. Eng and David Kazanjian (Berkeley: University of California Press, 2003), 467.

[6] Butler, "Afterword: After Loss, What Then?," 468.

[7] Butler, "Afterword: After Loss, What Then?," 468.

Grover Cleveland urging him to "strike the first steady blow toward lifting this burden of infamy from our country."[8]

Despite her persistence, the book – or any other of her endeavors – did not manage to change the situation; in fact, rather ironically, it promoted a vision of a romanticized past, with Franciscan missions and Mexican haciendas providing a backdrop to the amorous involvement of Ramona and Alessandro. At one point in the novel, one character embarks on a journey through various Indian settlements, and his experience seems to mirror Jackson's: "At each of these places he found Indians; miserable, half-starved creatures, most of them. … The ruins of the old Mission buildings were sad to see, but the human ruins were sadder."[9] The Native Americans, no longer "protected" under the old societal structures, chased away from their lands and left destitute, are monuments to a bygone era. They are not even human: they are ruins, like buildings; and like the edifices, their belonging to the Franciscans is never questioned. Likened to ruins, the Indians play the role of a reminder of the past, a mnemonic prompt to activate our memory.

In *Ramona,* the eponymous character, a girl of mixed origins, is also a reminder of a different kind. Her beauty, purity of heart and virtue, stressed time and again in the novel, were meant to move the hearts and conscience of the readers. Perhaps to make it easier for them to identify with Ramona's plight, the author gave her steely blue eyes. Her lover, Alessandro, on the other hand, bears an Italian, not a Native American or even Spanish name. Both, then, are "whitened" and thus, made palatable to a contemporary reader.

Such "whitening" also signals to the reader that race is not necessarily connected with superficial markers such as eyes or skin color. The moment Ramona falls in love with Alessandro, and thus the moment the main plot is set into motion, is also the moment of racial recognition and simultaneously, racial denial: "For the first time, she looked at him with no thought of his being an Indian, a thought there had surely been no need of her having, since his skin was not a shade darker than Felipe's; but so strong was the race feeling, that never till that moment had she forgotten it."[10] What the "race feeling" might be is never explained; the narrator assumes that no clarification is necessary. For a modern reader, however, it remains problematic: whose feeling is it? Whether it is what others feel

---

[8] Hunt Jackson, "A letter to President Grover Cleveland," *The New York Times* (September 29, 1885), <http://query.nytimes.com/mem/archive-free/pdf?res=F70E10FF355D10738DDDA00A94D1405B8584F0D3>.

[9] Hunt Jackson, *Ramona* (New York: Signet Classic, 2002), 213.

[10] Hunt Jackson, *Ramona,* 52.

when approaching Native American characters, or whether it is a manner that distinguishes them is an unresolved issue. Whatever the case, the narrator is at pains to present race not as a skin-deep problem: it is not even a matter of the way of living. Alessandro thinks "in his poetic Indian fashion,"[11] and if race is a way of thinking, as suggested, then it is necessarily and inevitably a matter of the past. Both Alessandro and Ramona are doomed. The future belongs to the energetic, business-like people who have no patience for poetic thinking.

In Senora Moreno, Ramona's custodian, Jackson creates a character that hates Ramona precisely for who she is: a mix. In Senora Moreno's perspective, Ramona as a character is a warning against the dangers of miscegenation. Senora Moreno explains the reasons for her loathing: "'If the child [i.e., Ramona] were pure Indian, I would like it better,' she said. 'I like not these crosses. It is the worst, and not the best of each, that remains.'"[12] A reader can sense an attempt at a logical explanation of the fear of racial confusion, one that hints at a vision of an ideal society, in which the best characteristics of the people are distilled to their pure form – the kind of society one might suspect Frederick Jackson Turner longed for.

It is not only Senora Moreno who rationalizes her abhorrence; the narrator also finds reasons for the woman's feelings: "No wonder that Senora Moreno never told the story. No wonder, perhaps, that she never loved the child. It was a sad legacy, indissolubly linked with memories which had in them nothing but bitterness, shame, and sorrow from first to last."[13] The woman, Ramona's legal guardian, never tells her of her mixed origin, as the story of her people, the Native Americans, is a story white people (with whom Senora Moreno obviously identifies) would conveniently like to forget. Senora Moreno's identification with whites is validated by the narrator, who describes her as "the restless, passionate, proud, sad heart,"[14] which situates her on the side of the people who are active, energetic, prideful, yet not always in the right (signified by "sadness" here): that is, white people. As problematic as such an identification is, it is strengthened through her complete rejection of Ramona as soon as the girl embraces her Native American heritage.

For Ramona, learning of her origins is likened to a "revelation," and it is not without consequence as it is "pointing out the path in which destiny

---

[11] Hunt Jackson, *Ramona,* 74.
[12] Hunt Jackson, *Ramona,* 22.
[13] Hunt Jackson, *Ramona,* 23.
[14] Hunt Jackson, *Ramona,* 183.

called her to walk."[15] Ramona chooses her destiny, because she cannot choose any other. The idea that one's descent determines one's fate explains not only personal choices, but also, and more importantly, justifies larger historical processes. In such a perspective, "blood," "heritage," "origin," or "race" are inescapable and their imprint has inevitable results. For white people, their destiny is expansion and westward progress; for the aboriginal population, the prospect is much bleaker.

This particular fatalism is all the more surprising in a novel ostensibly promoting Native American rights. To add to the ambiguity in the treatment of the central theme, Native Americans are, at various points, called "the poor creatures,"[16] "beasts,"[17] or "silly things,"[18] depending on how sympathetic to their cause is the character voicing these sentiments. In perfect concord with the parlance of the era, they are even called "devils"; and that in a sentence evoking the romanticized past of the missions: "it is a good thing for those poor Indian devils to get a bit of religion now and then."[19] Quite predictably, the fatalism of race is combined with a religious outlook, and in *Ramona* the two define the eponymous character.

When Ramona follows her fate through and joins Alessandro, the couple must look for a safe refuge, as Alessandro's tribe is stripped of their lands. The hardship of such a life does not perturb Ramona at first: "One must be always serious and not unhappy, but never too glad, I think, when he lives with nothing between him and the sky, and the saints can see him every minute."[20] Such a way of living means a close proximity to heaven, but its sterility is apparent, as too much delight is not prescribed. As soon as Ramona decides to fulfill her racial destiny, she displays features which stereotypically characterize Native Americans: her stoicism and appreciation of nature become all the more evident, especially when set against Alessandro's traits ("the tireless caution and infinite patience of his Indian blood").[21] Characteristically, she asks her lover: "I cannot believe that it is but two days I have lived in the air, Alessandro. This seems to me the first home I have ever had. Is it because I am Indian,

---

[15] Hunt Jackson, *Ramona,* 108.
[16] Hunt Jackson, *Ramona,* 15.
[17] Hunt Jackson, *Ramona,* 48.
[18] Hunt Jackson, *Ramona,* 200.
[19] Hunt Jackson, *Ramona,* 7.
[20] Hunt Jackson, *Ramona,* 138.
[21] Hunt Jackson, *Ramona,* 82.

Alessandro, that it gives me such joy?"[22] Even though the question remains unanswered, the reader is invited to think about the possibility that for Native American characters, who themselves resemble ruins, the air is their element and their proper dwelling place.

Belief in the afterlife, grounded in religious sentiment, is in fact, the only hope for the Native Americans. According to the narrative, just like the Franciscan priests and missionaries, they are all alike, dispossessed by the Americans. The main church representative in *Ramona,* Father Salvierderra, thus explains the common fate of the Native Americans and the missionaries of the region at the hands of the American newcomers: "We are all alike helpless in their hands, Alessandro. They possess the country, and can make what laws they please. We can only say, 'God's will be done,' and he crossed himself devoutly, repeating the words twice."[23] Now it is not only the fulfillment of racial destiny: the fate of the Native Americans, as well as the priests, it is a matter of God's plan for humanity.

Such an estimation of America's place in Providence's 'grand' design is not exceptional: John O'Sullivan outlines the country's prospects, speaking with confidence:

> The far-reaching, the boundless future will be the era of American greatness. In its magnificent domain of space and time, the nation of many nations is destined to manifest to mankind the excellence of divine principles; to establish on earth the noblest temple ever dedicated to the worship of the Most High – the Sacred and the True. Its floor shall be a hemisphere – its roof the firmament of the star-studded heavens, and its congregation an Union of many Republics, comprising hundreds of happy millions, calling, owning no man master, but governed by God's natural and moral law of equality, the law of brotherhood – of 'peace and good will amongst men.'[24]

The future he is sketching for America is a fulfillment of God's plan. The country itself becomes a more ideal sphere, a realization of the development of the nations of the world throughout the ages. Sweeping and somewhat totalitarian in its tone, the vision of America's glorious future is an execution of the sacred fate of the chosen nation. The ideals of the American Revolution are also there: equality, freedom, and brotherhood are the foundations upon which God's plan will be

---

[22] Hunt Jackson, *Ramona,* 138.

[23] Hunt Jackson, *Ramona,* 46.

[24] John O'Sullivan, "The Great Nation of Futurity," in *The United States Democratic Review*, vol. 0006 Issue 23, 427.

implemented for the good of all humankind. Just as is the case with futuristic visions, O'Sullivan's is limited by what was considered possible in American politics at that time: he does not find space for Native Americans, or any non-white persons; nor is he concerned with any potential "laws of sisterhood," to paraphrase his own statement.

In O'Sullivan's text, God sanctions American expansion; in Jackson Turner's more scientific assessment, it is the characteristics of the American people that at the same time justify and enable their push westwards. In *Ramona,* we have a combination of both: as one Native American character explains in an accusatory statement, "they [Americans] have made a law which will take the land from Indians; from us who have owned it longer than any can remember; land that our ancestors are buried in, – will take that land and give it to themselves, and say it is theirs."[25] What is stressed here is not only that Americans themselves make laws that authorize their actions – and such autonomy goes against the belief in the mandate from heavenly powers. What is also represented is the belief that Indian ownership of the land extends beyond memory. The memory of how long they have had the land is obliterated. The suggestion is that the laws of ownership are forgotten, just like the Native Americans themselves will be. Their ownership is validated through the fact that the land is marked by the graves of the ancestors, which clearly points to the past. There is no future for the Native American tribes.

The narrator of *Ramona* is aware of the injustice as much as she is of the irony behind the Natives' identification with nature: "The Temecula Indians had disappeared, that was all there was of it, – disappeared, like any wild creatures, foxes or coyotes, hunted down, driven out; the valley was rid of them."[26] The suggestion that appears here is that the land itself is purged of the inferior inhabitants and prepared for the future greatness of the white civilization. Hunt Jackson's narrator mourns the disappearance of the race and represents it with sympathy and yet, melancholically.

In Hunt Jackson's novel, a minor character, young Merrill, is an exposition of the nature of a frontiersman. The narrator uses a geological metaphor to describe his qualities:

> The character of the Western frontiersman is often a singular accumulation of such strata, – the training and beliefs of his earliest days overlain by successions of unrelated and violent experiences, like geological deposits.

[25] Hunt Jackson, *Ramona,* 170.
[26] Hunt Jackson, *Ramona,* 177.

> Underneath the exterior crust of the most hardened and ruffianly nature often remains – its forms not yet quite fossilized – a realm full of the devout customs, doctrines, religious influences, which the boy knew, and the man remembers.[27]

It is interesting to see that the narrator talks of the nature of frontiersmen in terms of natural phenomena; in contrast to Native Americans, who were compared to ruins, and thus situated in the realm of civilization – a bygone one, yet a civilization nevertheless. A subtle denial of the stereotypical representation of Indians as "children of nature," it is also a suggestion that the influx of the white newcomers to the frontier is governed by the laws of the natural environment, just as, we might suspect, the passing away of the Indian tribes is.

The character of the frontiersman in *Ramona* is presented as layers of sediment: an early layer, buried under later experiences of roughing it on the frontier, can be reduced to Christian upbringing. The narrative, after all, is a hopeful one: it advocates common humanity and compassion to others stemming from Christian sentiments. The young frontiersman, similar to other minor characters in the novel, is rid of his prejudiced views after he comes in contact with Native Americans and realizes their shared basic human condition. A comparison with Harriet Beecher Stowe's *Uncle Tom's Cabin* is in place here, as both novels advocated the rights of dispossessed groups, and both did so with reference to Christian ethics. What Hunt Jackson aims at is similar to Beecher Stowe's intention: to activate Christian feeling in her readers in order to change the disgraceful reality of the day.

The narrator in *Ramona* suggests that it is not easy for a hardened frontiersman to reach down to the deep layers of sympathy, yet it is not impossible: they resurface "[b]y sudden upheaval, in some great catastrophe or struggle in his mature life."[28] Just like in an earthquake or a landslide, the deeper residue is revealed: "Assembly Catechism definitions, which he learned in his childhood, and has not thought of since, ring in his ears," and the narrator adds that the process is by no means a straightforward one: "he is thrown into all manner of confusions and inconsistencies of feeling and speech by this clashing of the old and new man within him."[29] Just like the only hope of the Franciscan priests and Native Americans is to seek solace in religion, the narrative also hopes

---

[27] Hunt Jackson, *Ramona,* 226.
[28] Hunt Jackson, *Ramona,* 226.
[29] Hunt Jackson, *Ramona,* 226.

that in this confusion the old values, learnt before the life as a frontiersman, will prevail.

Hunt Jackson reaches deep; for Jackson Turner, in his frontier thesis, formulated a mere decade later, it is the frontier that runs deep in American national character. There is no place for sentiments; American nature may not be artistic, but it is sturdy and resilient:

> From the conditions of frontier life came intellectual traits of profound importance. ... The result is that to the frontier the American intellect owes its striking characteristics. That coarseness and strength combined with acuteness and inquisitiveness; that practical, inventive turn of mind, quick to find expedients; that masterful grasp of material things, lacking in the artistic but powerful to effect great ends; that restless, nervous energy; that dominant individualism, working for good and for evil, and withal that buoyancy and exuberance which comes with freedom – these are traits of the frontier, or traits called out elsewhere because of the existence of the frontier.[30]

The frontier is at the base now: it justifies the need for expansion and denies any deeper level of sympathy or remorse. The stress is on the material and practical aspects of American identity. Individualism, energy, restlessness – all growing out of the conditions of frontier life – are combined with moral ambiguity, as these traits are used for good as much as for evil. Such a vision of identity also suggests that the past is immaterial; it is the future that counts.

No culpability is admitted, and yet, in the denial of memory and the past there lie the unacknowledged layers of the American character. They are the dark underside of the rampant expansionism and they constitute what Butler calls "the loss of loss itself." The disavowed, un-mourned elements of American identity return hauntingly: there are four film adaptations of the story of Ramona, made between 1910 and 1936; a stage version is named the official play of the state of California. They testify to the need of exorcising the denied aspects of American history as much as they prove the need to remember. Even the most painful history seems more desirable than no history at all.

---

[30] Jackson Turner, "The Significance of the Frontier in American History," 15.

References

Butler, J., 2003, "Afterword. "After Loss, What Then?,"" in Eng, D. L., and Kazanjian, D., eds., *Loss: The Politics of Mourning,* Berkeley: University of California Press, 467-473.

Hunt Jackson, H., 2002 (1884), *Ramona,* New York: Signet Classic.

—. 1885, "A letter to President Grover Cleveland," *The New York Times,* <http://query.nytimes.com/mem/archive-free/pdf?res=F70E10FF355D10738DDDA00A94D1405B8584F0D3>.

Noble, D. W., 1985, *The End of American History: Democracy, Capitalism, and the Metaphor of Two Worlds in Anglo-American Historical Writing, 1880-1980,* Minneapolis: University of Minnesota Press.

O'Sullivan, J., 1839, "The Great Nation of Futurity," *The United States Democratic Review*, vol. 0006 Issue 23, 426-430.

Ridge, M., 1991, "The Life of an Idea: Frederick Jackson Turner's Frontier Thesis," *Montana: the Magazine of Western History,* vol. 41 no. 1, 2-13.

Turner, J. F., 1921 (1893), "The Significance of the Frontier in American History," in *The Frontier in American History,* New York: Henry Holt and Company, 1-38.

CHAPTER FIVE

JOURNEYS THROUGH HISTORY: TIME, SPACE, AND BODY IN ANDREA HAIRSTON'S "GRIOTS OF THE GALAXY"

AGNIESZKA PODRUCZNA

In the introduction to *So Long Been Dreaming*, an anthology of postcolonial speculative fiction, Nalo Hopkinson, the co-editor of the volume, states:

> Arguably, one of the most familiar memes of science fiction is that of going to foreign countries and colonizing the natives, and as I've said elsewhere, for many of us, that's not a thrilling adventure story; it's non-fiction, and we are on the wrong side of the strange-looking ship that appears out of nowhere.[1]

This remark manages not only to epitomise some of the core ideas of the genre – namely, to *boldly go where no man has gone before* – but also to establish the relationship between the genre and the colonial ideals of imperial expansion and conquest, in which it has been rooted since its inception. It would be possible, therefore, to see this quote as a reflection on one of the most important characteristics of speculative fiction – one which is often overlooked in favour of approaching this genre of literature as a purely escapist phenomenon – that is, the fact that speculative fiction serves as more of a mirror image of our world than an imaginary vision of a universe completely divorced from any measure of broadly understood reality. And indeed, as Alan Clarke remarks: "Many writers and critics

---

[1] Nalo Hopkinson, "Introduction," in *So Long Been Dreaming: Postcolonial Science Fiction and Fantasy*, ed. Nalo Hopkinson and Uppinder Mehan (Vancouver: Arsenal Pulp Press, 2004), 7.

have noticed that sci-fi can tell us more about the times in which they were produced than they can about any future they depict."[2] Similarly, in his definition of the genre, Damien Broderick emphasises the metaphorical and metonymic properties of science fiction, which he sees, as Adam Roberts puts it, "as an open-ended cultural phenomenon, which is particularly good at reflecting times of great cultural and technological change, of which our present age is a good example."[3] Given the fact that this type of fiction is – as already mentioned – deeply rooted in the colonial ideals of imperial expansion and conquest, it would be possible to argue that the phenomenon of postcolonial speculative fiction has emerged as a reaction to the essentially colonialist roots of the genre,[4] a certain sort of counter-discourse, which, as Helen Tiffin observes,

> is dynamic, not static: it does not seek to subvert the dominant with a view to taking its place, but, in Wilson Harris's formulation, to evolve textual strategies which continually 'consume' their 'own biases' (Harris 1985:127) at the same time as they expose and erode those of the dominant discourse ...[5]

Thus, it would be possible to see postcolonial speculative fiction as a genre which transcends the binary nature of the hegemonic imperial system and questions the most fundamental assumptions of the colonial discourse, while giving a voice back to the silenced colonial subject, allowing them, in the words of Nalo Hopkinson, to "build [them] a house of [their] own."[6]

The theme of journey, then, becomes as crucial for the attempts at reclaiming the discourse of science fiction by the hitherto silenced voices as it has been for the mainstream, which constructed the binary Self-Other, human-alien dichotomy in a world in which the manifest destiny of the human race has been extended to include the vast expanse of the known

---

[2] Alan Clarke, "Past Tense: History, Heritage and Ideology," in *Britishness and Cultural Studies*, ed. Krzysztof Knauer and Simon Murray (Katowice: Wydawnictwo Śląsk, 2000), 70.

[3] Adam Roberts, "Defining Science Fiction," in *Science Fiction. The New Critical Idiom* (London and New York: Routledge, 2000), 13.

[4] Michelle Reid, "Postcolonialism," in *The Routledge Companion to Science Fiction*, ed. Mark Bould, Andrew M. Butler, Adam Roberts, and Sherryl Vint (London and New York: Routledge, 2009), 257.

[5] Hellen Tiffin, "Post-Colonial Literatures and Counter-Discourse," in *The Post-Colonial Studies Reader*, ed. Bill Ashcroft, Gareth Griffiths, and Helen Tiffin (London and New York: Routledge, 2002), 96.

[6] Hopkinson, "Introduction," 8.

universe. But for some postcolonial science fiction writers, it seems, the journey does not exist primarily in the spatial sense – instead, what they appear to be most interested in is its temporal aspect. The characteristics of the genre – one which facilitates the possibility for the postcolonial practice of writing back – allow those authors to construct stories of journeys not only through space, but also through time, which, in turn, enables a connection with the lost ancestral heritage, which was obliterated or misappropriated in the process of colonisation. In "Griots of the Galaxy," Andrea Hairston proposes a simultaneous movement forward (through space, from one body to another) and backward (through time), signifying the practices of postcolonial science fiction writing, which at the same time turns to the past in order to re-examine the lingering colonial tensions from a new angle and looks ahead to imagine different futures for those who had been denied the possibility to speak by the hegemonic discourse. Ultimately, then, what the journey in Hairston's story stands for is a metaphor for reclaiming and reimagining the past, for rewriting history from the perspective of the marginalised subject, for the process of postcolonial writing back.

In the epigraph to the story, Hairston writes: "The Griots of West Africa are musicians, oral historians, praise singers negotiating community. They stand between us and cultural amnesia. Through them we learn to hear beyond our time and understand the future."[7] The griots in Hairston's story serve a similar purpose: they are a race of alien beings who possess the ability to inhabit bodies of dying or recently deceased beings – be they people, other animals or plants – in order to gather knowledge of their history and collect their experiences so that they can be recorded for posterity among the stars, quite literally building the futures of all the species out of their past experiences. Their relationship with time is, therefore, strongly connected with their continuous journey from body to body, since, in the act of re-living and relaying the memories of the newly inhabited bodies, the griots are capable of grasping the past, the present and – in a way – the future in one single moment. As the protagonist, Axala, remarks herself:

> Body historians, griots of the galaxy, we didn't diddle ourselves in jungle paradises, we inhabited flesh to gather a genealogy of life. We sought the story behind all stories. Collecting life's dazzling permutations, however

[7]Andrea Hairston, "Griots of the Galaxy," in *So Long Been Dreaming: Postcolonial Science Fiction and Fantasy*, ed. Nalo Hopkinson and Uppinder Mehan (Vancouver: Arsenal Pulp Press, 2004), 23.

sweet or sour, was our science, religion and art – nothing nobler in eternity.[8]

The objective of this practice, however, is not to reach one grand, unifying narrative; quite the opposite, the story behind all stories mentioned by the protagonist consists of polyphonic voices and disjointed, fragmented histories which, in Bakhtinian terms, challenge the seemingly absolute forms of authority and their claims to universality, and which reflect the reality of the postcolonial experience, with its ruptures in meanings and fractured narratives. In the story, the idea of resisting the colonial master narrative manifests itself in the way in which the griots remember their past borrowed lives – they are referred to as serial amnesiacs, as they can see only what Hairston calls the Edges, disjointed fragments of their former selves, and it is only after they upload their memories to the griots' repository of knowledge that those histories are allowed to unfold in their entirety. This, once again, emphasises the peculiar relationship that the griots have with the notion of time, in which past, present and future progress from the linear structure of time to a model in which those three are allowed to exist simultaneously, represented by the memories, the re-lived experience of the griots and the future unfolding of the collected histories respectively.

At the same time, the story establishes the body as a place of history, and, consequently, as a place of journey for the griots, who travel from one life to another in order to tell the stories that otherwise would be forgotten. Just like the griots of West Africa, referred to by the author in the epigraph to the story, Axala's people collect stories and histories that preserve the past and safeguard the future, drawing directly from the source of the lived and re-lived experience – the physical body. For Hairston, then, the bodies inhabited by the griots become excavation sites, places from which the process of writing back is enabled. Similarly to several authors who also use the imagery of the body (especially that of a colonial subject) as a repository of ancestral heritage, such as Celu Amberstone or Suzette Mayr, Hairston emphasises the link between the corporeal and the historical, creating a narrative in which the body, as a carrier of memories, facilitates the rediscovery of stories and voices, and stands as a metaphor for the process of reclaiming, letting the subaltern speak. At one point the protagonist remarks, "You didn't get to choose a life; you only had minutes to find what was available. A true historian should be glad for any dying body to ride. Every story offered precious insight."[9] This journey

---

[8] Hairston, "Griots of the Galaxy," 27.
[9] Hairston, "Griots of the Galaxy," 25.

from body to body, from life to life is, therefore, at the same time, a journey through time, a journey of discovery of that which had been erased, forgotten, and buried alongside the extinguished memories that Albert Memmi mentions in *The Colonizer and the Colonized*[10] as the inevitable result of the presence of the hegemonic colonial discourse. The metaphor of colonial erasure and appropriation is further explored in the context of rape which Renee – the original owner of the body that the protagonist currently inhabits – had experienced in the past and of which she has initially no memory. The way in which Hairston structures this particular part of the narrative seems to reflect the horrific violations brought upon the colonised by the colonisers that resulted in an almost complete erasure of the indigenous histories and the ultimate extinguishing of memories. It is only the act of drawing from the past in the journey of rediscovery that allows Axala/Renee to confront her trauma and move forward, much like – Hairston seems to suggest – the journey through history to uncover silenced voices and lost histories, which permits the voices of the Othered colonial subjects to be heard, allows for a re-examination of the lingering colonial traumas and tensions that haunt the indigenous communities.

In this way, the journey through history presented by Hairston seems to represent a sort of subversive archaeology, which addresses the discourse that has been since its beginnings oriented towards the central rather than the peripheral, and which could be characterised by its reification of the binary Self-Other dichotomy imposed by the colonial order, manifesting in the act of looking from the outside in. As Matthew Liebmann remarks in the introduction to *Archaeology and the Postcolonial Critique*, "[f]rom the earliest days of the discipline, archaeology has played a part in creating and controlling the representation of the past in colonized societies."[11] In the story, however, the control over the unearthing of the past and, consequently, the control over the discourse is handed over to the griots, who would be coded by the mainstream as the alien and, therefore, the Other. By allowing the Othered colonial subjects to speak – as the "[g]riots were storytellers, no matter the story"[12] – the author establishes their journey from body to body in search of lost memories and almost forgotten experiences as a process of the rewriting of history from the

---

[10] Albert Memmi, *The Colonizer and the Colonized* (Boston: Beacon Press, 1993), 52.

[11] Matthew Liebmann, "Introduction: The Intersections of Archaeology and Postcolonial Studies," in *Archaeology and the Postcolonial Critique*, ed. Matthew Liebmann and Uzma Z. Rizvi (Lanham: AltaMira Press, 2008), 6.

[12] Hairston, "Griots of the Galaxy," 30.

point of view of the marginalised subjects, by filling in the gaps left by the colonial attempts at erasure and eschewing the colonial master narrative in favour of polyphonic, fragmented stories and ruptured narratives that reflect the disjointed, complex reality of the colonial experience.

This subversive practice, which comes from the body understood both as a locus of colonisation and as a place of ancestral memory, grounded in blood, flesh and remembrance, is facilitated by a journey through history – or rather histories, as history becomes here a non-homogenous polyphony of fragmented narratives – undertaken by the griots, who protect those silenced voices from total obliteration, but at the same time, this journey becomes also a journey of self-discovery, mirroring the processes of creation of the subject in the colonial and postcolonial paradigm. Similarly to other works of postcolonial speculative fiction, the creation of the subject's identity is closely linked to the body itself as well as the embodied ancestral heritage, and results in the emergence of hybrid identities, which manifest themselves in the text in the figures of the griots. This, in turn, allows the author to write against the essentialism inherent in the Self-Other dichotomy as well as any claims to so-called authenticity of the colonial indigenous experience and create a Third Space from which the silenced colonial subject can speak.

But what Hairston stresses the most throughout the whole narrative is the importance of the journey both through time and through lives, a journey which simultaneously creates the future and preserves the past in the present moment, thus entering the pattern of constant change and continuity. At one point in the story, the protagonist says, "I was caught in deep time, with our ancestors walking through my body, making sense of the present moment from so many gone by."[13] It becomes evident, then, that the journey of discovery conceived of only in the spatial sense would ultimately prove to be insufficient for the processes facilitating the creation of the subject, since, as Hairston suggests, it is possible for those who had been for the longest time marginalised and denied the possibility to speak to make sense of their present and their future only in relation to their past. The fact that the griots are capable of bridging the gap between the past, the present, and the future allows them to become the ultimate solution to the experiences of colonial erasure, but, at the same time, it affects their own identities. "Each body changes us. We are the sum of all the bodies we have joined," says the narrator at one point. "I am not the same Axala that you were."[14] In the process of jumping from body to

[13] Hairston, "Griots of the Galaxy," 40.
[14] Hairston, "Griots of the Galaxy," 45.

body, from life to life, from one story to another, the griots become, therefore, irrevocably changed. This, on the one hand, reflects the way in which the unearthing of the forgotten past facilitates a deeper connection with the ancestral heritage of the subjects, and, on the other hand, it mirrors the changes experienced by the Othered subjects, resulting from the presence of the colonial system, such as Bhabha's mimicry or hybridity. Thus, it seems that Hairston writes against the essentialist demands for ethnic authenticity in the postcolonial space of difference, and agrees with Helen Tiffin, who argues that "such pre-colonial cultural purity can never be fully recovered."[15] What Hairston stresses, instead, is what she calls the "polyphony of memories"[16] that characterises the postcolonial condition. This, in turn, emphasises the importance of history, collective memory and the oral tradition of storytelling, which allow the subjects to construct themselves in relation to their past and make sense of their place within their community in the present as well as safeguard their future against total obliteration. In order to illustrate this point, towards the end of the story, Hairston writes:

> I was in the home grove, the bosom of the family, connected to deep time, praise singing life. All of our griot experiences, a polyphony of memories, rushed through me. Together the trees and I made meaning and broadcast twice ten thousand years, the incredible story of life on Earth to the stars. Imbedded in this polyrhythmic history, I remembered everything, the story I'd become. Axala of Earth.[17]

In the end, what Hairston presents the readers with is an account of the subversive strategies which allow the subjects in postcolonial realities to construct their own identities and reclaim stories which were appropriated and silenced by the colonial enterprise. Facilitated by the journey into the self as well as the journey into history, those archaeologies of the past fill in the gaps left by the processes of erasure and bridge the multiplicity of diverse ethnic experiences while allowing them to remain polyphonic in nature. Hairston's non-linear approach to the notion of time as well as the insistence upon its significance in the processes of re-living, remembering, reclaiming, and, ultimately, the postcolonial practice of writing back are particularly emphasised in the peculiar relationship between the griots and history (or rather: the multiplicity of histories) as well as the future – a relationship which represents the postcolonial attempts at reconciliation

[15] Tiffin, "Post-Colonial Literatures and Counter-Discourse," 95.
[16] Hairston, "Griots of the Galaxy," 38.
[17] Hairston, "Griots of the Galaxy," 38.

with the legacy of the imperial experience and which prompts a re-examination of the Othered self and, ultimately, leads to the reclamation of the discourse which denied the colonial subjects the possibility to speak. Therefore, it is only once the former colonial subjects, whose diasporic existence is embodied in the figures of the griots, can reconcile their past with their present, and once they can re-examine and rewrite their own histories, drawing from the repository of their ancestral heritage, that their voice may finally be heard. The journey through time, then, becomes the catalyst of change – it allows for a transformation within the colonial dynamics of difference, embodied in their initial state by the binary colonial ideal of the Self-Other dichotomy, and facilitates the process of the creation of a new, postcolonial identity for the formerly Othered subjects. In this process, the movement downwards, towards the hidden ancient repositories of ancestral knowledge, represented by the metaphor of the unearthing of the past (in the form of the re-lived memories), is accompanied by the simultaneous movement backwards and forwards through time, symbolising the complex nature of the journey that the griots of the galaxy undertake in order to preserve the voices that would otherwise be forgotten and gather "a genealogy of life."[18] And this, according to Hairston, is where their journey ultimately leads them: "After twice ten thousand years I wanted to do something impossible, something noble," Axala says in the final paragraph of the story. "Instead of chasing down infinity, we could contribute our souls to Earth. A blessing on this future, not now or nothing. The voice and the body and the history."[19]

References

Clarke, A., 2000, "Past Tense: History, Heritage and Ideology," in Knauer, K., and Murray, S., eds., *Britishness and Cultural Studies*, Katowice: Wydawnictwo Śląsk, 70-82.

Hairston, A., 2004, "Griots of the Galaxy," in Hopkinson, N., and Uppinder Mehan, eds., *So Long Been Dreaming: Postcolonial Science Fiction and Fantasy*, Vancouver: Arsenal Pulp Press, 23-45.

Hopkinson, N., 2004, "Introduction," in Hopkinson, N., and Uppinder Mehan, eds., *So Long Been Dreaming: Postcolonial Science Fiction and Fantasy*, Vancouver: Arsenal Pulp Press, 7-9.

Liebmann, M., 2008, "Introduction: The Intersections of Archaeology and Postcolonial Studies," in Liebmann, M., and Rizvi, U. Z., eds.,

[18] Hairston, "Griots of the Galaxy," 27.
[19] Hairston, "Griots of the Galaxy," 38.

Archaeology and the Postcolonial Critique, Lanham: AltaMira Press, 1-20.

Memmi, A., 1993, *The Colonizer and the Colonized*, Boston: Beacon Press.

Reid, M., 2009, "Postcolonialism," in Bould, M., Butler, A. M., Roberts, A., and Sherryl Vint, eds., *The Routledge Companion to Science Fiction*, London and New York: Routledge, 256-266.

Roberts, A., 2000, "Defining Science Fiction," in Roberts, A., ed., *Science Fiction. The New Critical Idiom*, London and New York: Routledge, 1-46.

Tiffin, H., 2002, "Post-Colonial Literatures and Counter-Discourse," in Ashcroft, B., Griffiths, G., and Tiffin, H., eds., *The Post-Colonial Studies Reader*, London and New York: Routledge, 95-98.

CHAPTER SIX

TRAUMATIC AND NARRATIVE TIME IN GRAHAM SWIFT'S *WATERLAND*

SŁAWOMIR KONKOL

Waterland (1983), Graham Swift's undeniably still most widely known and most critically acclaimed novel, is a text preoccupied with time. Deserving – perhaps with certain reservations – the name of a historical novel, famously interweaving various methods of describing temporality, different ways of recording the past, it has appropriately been quoted as a textbook example of "historiographic metafiction."[1] Private and public histories, scientific and political accounts, fairy tales and chronicles, all ultimately circle around the indescribable trauma of "the Here and Now," the immediate reality which in the end always eludes verbalisation. Self-referential to a high degree, *Waterland* is also a narrative of narrativisation, of overcoming the time-stopping traumatic moment by incorporating it into the symbolic order of language, which provides a shelter from the life-threatening immediacy of experience. At the same time, it is a novel of lost innocence, of entry into the structures of symbolic exchange seen as a banishment from Eden. These two approaches, in fact, define at least some of the notions of time in the text, ones which will be crucial to my argument. The first of these finds expression in images invoked by the narrator of his idyllic pre-symbolic existence, of "prehistorical, pubescent times, when we drifted instinctively without the need for prior arrangement, to our meeting place,"[2] of times before sexuality and the imaginary-symbolic reality of adulthood, of times of unity with the circular, timeless natural history. In this context, narratives appear to be a form of trickery by means of which the individual is tempted out of the

---

[1] The term was coined by Linda Hutcheon. In her *A Poetics of Postmodernism: History, Theory, Fiction* (London & New York: Routledge, 1988), she refers to *Waterland* directly at several points (see pages 15, 55, 108, 117).
[2] Graham Swift, *Waterland* (London: Picador, 1983), 44.

safe confines of an imagined wholeness, only to be trapped in the unstoppable passing of time. On the other hand, the confession of the protagonist, Tom Crick, is driven by the horrifying, traumatic temporality experienced by those who are suspended in "another world, where things go on happening,"[3] out of which the fabrication of narrative, by ordering chaotic experience, offers an escape, however tentative (its fragile construction may, after all, be disrupted at any moment). The divergent functions of narration in Tom Crick's story may, in fact, be said to replay the tension between Lacan's three orders. After all, storytelling has the function of a fantasy providing a comforting substitute for the desired, imaginary union with the mother and a protection against the terrifying contingency of the human condition, interrupting the symbolic order in its striving to displace the real. The sense of unity and wholeness is in this understanding a form of auto-narration, covering over the intrusions of trauma which disturb what might be regarded in the context of Lacan's notion of "future anterior,"[4] implying the retroactive and unending construction of the subject. In this sense, the subject is never fixed, always anticipated rather than confirmed and, for this reason, what is considered to be its past in fact sets out the trajectory of its potential future. As Lacan famously put it: "What is realised in my history is not the past definite of what was, since it is no more, or even the present perfect of what has been in what I am, but the future anterior of what I shall have been for what I am in the process of becoming."[5] Trauma disrupts this process, as an experience which happens "too soon" or "too fast" to be comprehended by the subject, with catastrophic consequences, or, to quote Sigmund Freud, "an experience which within a short period of time presents the mind with an increase of stimulus too powerful to be dealt with or worked off in the normal way."[6] For Lacan, the very constitution and functioning of a speaking subject is traumatic. As Elizabeth Wright puts it, speaking about the notion of a "missed encounter" with the inassimilable real, the attempt to overcome the invasion of the consciousness by the unconscious, the

---

3 Swift, *Waterland*, 263.

4 Matthew Sharpe "Lacan, Jacques," *Internet Encyclopedia of Philosophy*, < http://www.iep.utm.edu/lacweb/ >.

5 Jacques Lacan, *The Language of the Self: The Function of Language in Psychoanalysis*, trans. Anthony Wilden (Baltimore, Maryland: John Hopkins University Press, 1981), 63.

6 Sigmund Freud, "The Introductory Lectures on Psycho-Analysis," in *The Standard Edition of the Complete Psychological Works of Sigmund Freud*, ed. and trans. James Strachey (London: Hogarth Press and Institute of Psychoanalysis, 1963), vol. 16, 275.

"separation produced by the symbolic, particularly in the earliest moments, will break into the chaotic experience of the infant and produce shock. To 'miss' the encounter is to fail to grasp the brute experience. The symbolic works upon the real by means of repetition, but never turns into it."[7] The lack which lies at the core of the Lacanian subject is hidden in self-creation achieved through the coherence of a narrative, with language serving to tame the trauma by producing a self-image associated with Lacan's mirror stage and creating an illusion of the ideal self by repressing the real. On the other hand, however, storytelling also stands for the inadequacies and dissatisfaction associated with the symbolic order, the incessant substitution condemning the subject to an endless search for what can never possibly be found. After all, the point of the relation between the symbolic and the real lies in the tension between the subject's imagined identification with signifiers and the impossibility of achieving their adequate representation in language. Much as the rules of social interaction determined in the symbolic order protect the subject from the unbearable pressure of the real, they alienate him or her from anything that might be considered an "authentic self," initiating a futile striving to regain the presumed pre-Oedipal fullness. In the words of Paul Verhaeghe: "The interaction between the two levels consists in the never ending attempt of the chain of signifiers to produce an answer to the real. This attempt fails and results in the exact opposite: the more signifiers produced, the further one moves away from this real ... . The subject chooses the (m)Other in order to regain the lost paradise of the primary experience of satisfaction, and the net result will be an ever more clear delineation of this loss."[8]

In line with this impasse, much as in all of his other texts, Swift oscillates in *Waterland* between an embrace and a distrust of narrativisation as a technique of taming traumatic experience. Early on in the novel, its protagonist announces to his listeners: "Your history teacher wishes to give you the complete and final version."[9] Predictably, his story proves to be far from achieving anything like this goal. Instead, the achronological, meandering text, a collage of various genres and perspectives, constantly changing time planes and locations, reflects the impossibility of containing reality within an authoritative account. The narrative is centred around a traumatising series of events which it repeatedly tries to domesticate through storytelling, at the same time pondering on the value of this

[7] Elisabeth Wright, *Lacan and Postfeminism* (Cambridge: Icon Books, 2000), 41.
[8] Paul Verhaeghe, "Causation and Destitution of a Pre-ontological Non-entity: On the Lacanian Subject," in *Key Concepts of Lacanian Psychoanalysis*, ed. Dany Nobus (New York: State University of New York Press, 1988), 176.
[9] Swift, *Waterland*, 6.

strategy. Sean Homer's description of the goals central to Roland Barthes' *Camera Lucida* appears to be applicable to what Swift – or perhaps rather Tom Crick, his narrator – aims to achieve in the novel: "The text continually circles around this absence and produces a series of substitute [objects] that fill the hole left by the original loss, but we can never get back to that original experience itself. The text never produces, indeed it can never produce, the Truth, the thing-in-itself, the essence that [he] is searching for."[10] Taking a cue from the quoted author himself, Sean Homer uses Barthes' text to illustrate the relationship of the Lacanian categories of the real and the symbolic: as a reflection motivated by a photograph of his mother, impossible for him to retrieve, it reproduces the function of the lost object of psychoanalysis. In turn, this relationship may be seen more generally as emblematic for photography as a medium: "Unlike painting or language, photography can never deny its past, that the thing existed and was there in front of the camera, but that real is lost the moment the photograph itself comes into being. And it is this that is the very essence of photography – its *noeme*, that is to say, its 'that-has-been' or its intractability."[11] In Lacanian psychoanalysis, the orders of the real and the symbolic are, indeed, postulated to be in an analogous relationship, the two being interdependent and mutually exclusive. After all: "It is through the process of cancelling out, of symbolizing the real, that 'social reality' is created."[12] When the symbolic subject emerges, perceiving itself as distinct from the incomprehensible magma of the pre-symbolic existence, the real itself is replaced with the categories provided by the imaginary and symbolic. At the same time, the symbolic can never exhaust the material of the real, symbolise it in its entirety. Description never fully fits the experience, and whatever part of what is called "reality" remains unsymbolised (or indeed impossible to symbolise), is assumed to belong to the real. Trauma is one of its faces: a leftover of symbolisation which constitutes a blockage in the process of substitution.[13] In turn, the interventions of this traumatising remainder serve to undermine the stability of the imaginary-symbolic construct of reality, repeatedly bringing out its inevitable contingency. In Lacan's later teachings, the real is closely associated with the notion of *jouissance*, against which language is perceived to be a form of defence. In fact, according to Bruce Fink, "[t]he subject comes into being as a form of attraction toward and defence against a primordial, overwhelming experience of what the French call

[10] Sean Homer, *Jacques Lacan* (London: Routledge, 2005), 93.
[11] Homer, *Jacques Lacan*, 93.
[12] Homer, *Jacques Lacan*, 83.
[13] Homer, *Jacques Lacan*, 83.

jouissance: a pleasure that is excessive, leading to a sense of being overwhelmed or disgusted, yet simultaneously providing a source of fascination."[14] Dylan Evans notes another significant aspect of *jouissance*, i.e. its paradoxical and subversive nature: the defence against its overwhelming influence, the pleasure principle, implied by the very structure of language, prohibits access to *jouissance*,

> which is why '*jouissance* is forbidden to him who speaks, as such' (E, 319). The subject's entry into the symbolic is conditional upon a certain initial renunciation of *jouissance* in the castration complex, when the subject gives up his attempts to be the imaginary phallus for the mother; 'Castration means that *jouissance* must be refused so that it can be reached on the inverted ladder *(l'échelle renversée)* of the Law of desire' (E, 324). The symbolic prohibition of enjoyment in the Oedipus complex (the incest taboo) is thus, paradoxically, the prohibition of something which is already impossible; its function is therefore to sustain the neurotic illusion that enjoyment would be attainable if it were not forbidden. The very prohibition creates the desire to transgress it, and *jouissance* is therefore fundamentally transgressive (see S7, ch. 15).[15]

This is the reason why, for Lacan, the subject is, as a result of the symbolic castration, of abandoning the wish to be the imaginary phallus for the mother and of accepting the rules of symbolic exchange, deprived of direct contact with past experience, presumably captured in signifiers. Matthew Sharpe points out the analogy of this position "to the Kantian postulation that finite human subjects are debarred from immediate access to things in themselves."[16] This, in turn, implies that the original lost object, "which would as it were be 'more real' than the other things that the subject can readily signify … is what is primordially repressed when the subject accedes to becoming a speaking subject at castration. When the subject accedes to the symbolic, [Lacan] repeats, the Real of aspired-to incestuous union, and the sexualized transgressive enjoyment or *jouissance* it would afford, is necessarily debarred."[17]

The words of Tom Crick on the dangers of immediate experience capture both the dominance of the symbolic-imaginary construct of reality and its vulnerability to invasion by the real: "[T]he Here and Now, which

---

[14] Bruce Fink, *The Lacanian Subject. Between Language and Jouissance* (Princeton: Princeton University Press, 1995), xii.
[15] Dylan Evans, *An Introductory Dictionary of Lacanian Psychoanalysis* (London and New York: Routledge, 1996), 94.
[16] Sharpe, "Lacan, Jacques."
[17] Sharpe, "Lacan, Jacques."

brings both joy and terror, comes but rarely – does not come even when we call it. That's the way it is: life includes a lot of empty space. We are one-tenth living tissue, nine-tenths water; life is one-tenth Here and Now, nine-tenths a history lesson. For most of the time the Here and Now is neither now nor here."[18] On the other hand, the inclusion of "both joy and terror" in his description corresponds to the ambiguous status of *jouissance*, that which is at the same time infinitely tempting and repulsive or unbearable since "every drive is an attempt to go beyond the pleasure principle, to the realm of excess *jouissance* where enjoyment is experienced as suffering."[19] The allure of *jouissance* is grounded in the fact that the fulfilment of desire in the symbolic order of language relations is by definition impossible, that the satisfaction offered by the symbolic order will always prove insufficient. This frustration requires the support of fantasy to enable the subject to function within the symbolic:

> As we seek to realize our desires we will inevitably be disappointed – the satisfaction we achieve is never quite enough; we always have the sense that there is something more, something we have missed out on, something more we could have had. This something more that would satisfy and fulfil us beyond the meagre pleasure we experience is *jouissance* ... . This belief in the excessive *jouissance* of the Other is sustained through fantasy. Fantasy is one of the ways through which we reconcile ourselves to our dissatisfaction with our own *jouissance* and the impossibility of the real. Through fantasy we construct our social reality as an answer to the intractability of the real.[20]

Lacan's notion of fantasy is that of a reinforcing wall for the symbolic-imaginary construct of reality which protects the subject against the real invading his or her experience. Psychoanalytic treatment ideally culminates in what Lacan termed "traversing the fantasy," where the trauma of the real is to be subjectified. This means that "the subject takes the traumatic event upon him/herself and assumes responsibility for that *jouissance*,"[21] and recognises that the supposedly lost object being sought for is no more than a compensation for the trauma of castration. However, the traumatic intensity of the real is fundamentally insupportable and, therefore, "as Ellie Ragland-Sullivan puts it, human beings will settle for

---

[18] Swift, *Waterland*, 52.
[19] Evans, *An Introductory Dictionary of Lacanian Psychoanalysis*, 34.
[20] Homer, *Jacques Lacan*, 89-90.
[21] Homer, *Jacques Lacan*, 83.

any experience, however painful, rather than fall out of the familiarity of the symbolic into the trauma and void of the real."[22]

The plot of *Waterland* emphasises the contingency of the socially constructed reality by means of numerous episodes in which irruptions of chaos unsettle its structures time and time again. Repeatedly described as taking the human being, defined by Crick for the purposes of his harangue as "the story-telling animal," outside the flow of events of narrativised time, the notion of "the Here and Now" is beyond any doubt as central to his reflection as the category of the real ultimately became to Lacan's teachings. In an attempt to determine its nature for his listeners, Crick offers the following observations:

> But what is this much-adduced Here and Now? What is this indefinable zone between what is past and what is to come; this free and airy present tense in which we are always longing to take flight into the boundless future?
>
> How many times, children, do we enter the Here and Now? How many times does the Here and Now pay us visits? It comes so rarely that it is never what we imagine, and it is the Here and Now that turns out to be the fairytale, not History, whose substance is at least for ever determined and unchangeable. For the Here and Now has more than one face. It was the Here and Now which by the banks of the Hockwell Lode with Mary Metcalf unlocked for me realms of candour and rapture. But it was the Here and Now also which pinioned me with fear when livid-tinted blood, drawn by a boat-hook, appeared on Freddie Parr's right temple.[23]

The Here and Now is associated by the narrator both with the carnal rapture of his discovery of sexuality and with the nauseating realisation that because of this discovery and a tangled web of jealousy and lies, his childhood friend, Freddie, has been murdered by Tom's half-brother. Therefore, in Tom Crick's life story, the unexpected, menacing immediacy of events stretches literally between the extremes of the different conceptions and faces of *jouissance* in Lacanian psychoanalysis: from the biological dimension of satisfying a need and sexual pleasure through the terrifying consequences of overlooking the demands of the pleasure principle (that is, neglecting the costs of gaining one's pleasure),[24] the

---

[22] Quoted in Homer, *Jacques Lacan*, 89-90.

[23] Swift, *Waterland*, 51-2.

[24] See Evans, *An Introductory Dictionary of Lacanian Psychoanalysis*, 93-4. Since the 1960s, Lacan has elaborated in his theory the opposition of *jouissance* and pleasure, "an opposition which alludes to the Hegelian/Kojevian distinction between *Genuß* (enjoyment) and *Lust* (pleasure) (cf. Kojève, 1947:46). The

"feeling in the guts," piercing through the protection of symbolic constructs presumed to tame the chaos of the world around us, to the Nothing of unmediated reality, constantly invading on "fragile islands of life."[25] However, the affinity of Crick's Here and Now with the Lacanian real may ultimately appear debatable, if only because the notion of *jouissance* presumes an excess which is difficult to observe in all manifestations quoted by Crick to illustrate the notion. In fact – and more significantly – his own words about the Here and Now "turning out to be the fairytale" reveal the idealising manipulation to which he subjects the idea.

The face of the Here and Now that is probably the most prominent in *Waterland* – and unquestionably excessive in exploding the categories of a subjectifying experience – is trauma, the time-stopping, ceaseless repetition of the event which proves impossible to assimilate. The clearest example here is perhaps the hospital for veterans where the narrator's mother, Helen, used to work, struggling to re-incorporate into the structures of the symbolic order people frozen in the traumatic time by their experience of war. Among them was Tom's father, Henry Crick, who "finds it hard to separate in his mind the familiar-but-foreign fields of the Fens and the foreign-but-familiar mudscapes he has come from … . He thinks: there is only reality, there are no stories left. About his war experience he says: 'I remember nothing.'"[26] Eventually, with his future wife's assistance, he "learns, also, to tell those stories of old Flanders ... which will lead on to other stories, till the pain, save for sporadic twinges in the knee, is almost gone."[27] The introduction of narrative chronology allows one to put the experience in the past. The imposition of a time reference operates as a neutralising symbolic framework which enables the subject to function properly again. Significantly, it is also Helen Crick who exemplifies the limitations of this strategy, as another very potent illustration of the confrontation between narrative and trauma is constituted by Crick's memory of witnessing his mother's death. The same figure who laboured against the traumatising exclusions from the symbolic

pleasure principle functions as a limit to enjoyment; it is a law which commands the subject to 'enjoy as little as possible'. At the same time, the subject constantly attempts to transgress the prohibitions imposed on his enjoyment, to go 'beyond the pleasure principle'. However, the result of transgressing the pleasure principle is not more pleasure, but pain, since there is only a certain amount of pleasure that the subject can bear."

[25] Swift, *Waterland*, 296.

[26] Swift, *Waterland*, 17.

[27] Swift, *Waterland*, 194-195.

and who first introduced Crick himself to the power of storytelling now in turn introduces him to the dimension of reality which exceeds description and, in a sense, subverts the power of narration:

> She croaks, after a long pause in which it seems she will never quite find enough breath to speak, "Well this … is a fine … state of things … isn't it?" With an air of almost-levity. But her eyes say something different. They say: Look children, your mother's dying. In a little while she won't be here any more. It's a unique, a momentous event. Unique and momentous, not to say unexpected, for your mother too. It only happens once, it won't be repeated for you. Note it, observe it.
>
> (And I did, children, very carefully. And though, indeed, it only happened once, it's gone on happening, the way unique and momentous things do, for ever and ever, as long as there's a memory for them to happen in ...).[28]

The uniqueness of the traumatising event, however, is in *Waterland* perceived as an exception to the overwhelming awareness of the numbing nothingness of reality, its nauseating emptiness: storytelling is for Swift's characters a means of escape not only from the unexpectedly drastic encroachments of the chaos of the world into their fantasies but, primarily, from the dread of living in the liquid landscape of the Fens: "And what are the Fens, which so imitate in their levelness the natural disposition of water, but a landscape which, of all landscapes, most approximates to Nothing?"[29] The horrifying necessity of facing the faceless reality of the Fens is, typically for Swift, shown to be parallel to the personal plight of the narrator and the condition of the civilisation in which he lives. When his wife kidnaps a baby, claiming that the deed was ordered by God, Crick replies by reflecting on the bleak terror of God's silence or, indeed, absence: "But God doesn't talk any more. Didn't you know that, Mary? He stopped talking long ago. He doesn't even watch any more, up there in the sky. We've grown up now, and we don't need him any more, our Father in Heaven. We can fend for ourselves. He's left us alone to make what we will of the world … God's for simple, backward people in godforsaken places."[30] In an equally Swiftian move, all human endeavour – from storytelling to empire building – is further presented as part of our attempts to overcome this terrifying feeling: "Yes, the end of the world's on the cards again – maybe this time it's for real. But the feeling's not new. Saxon hermits felt it. They felt it when they built the pyramids to try to

[28] Swift, *Waterland*, 237-8.
[29] Swift, *Waterland*, 11.
[30] Swift, *Waterland*, 232-3.

prove it wasn't true. My father felt it in the mud at Ypres. My grandfather felt it and drowned it with his suicidal beer. Mary felt it ... It's the old, old feeling, that everything might amount to nothing."[31] In fact, as one of the possible alternatives of conceptualising history, Crick offers to his students the view that the reality is essentially devoid of any positive content and that history's recounting of a series of significant events is merely designed to divert our attention from this realisation:

> Reality's not strange, not unexpected. Reality doesn't reside in the sudden hallucination of events. Reality is uneventfulness, vacancy, flatness. Reality is that nothing happens. How many of the events of history have occurred, ask yourselves, for this and for that reason but for no other reason, fundamentally, than the desire to make things happen? I present to you History, the fabrication, the diversion, the reality-obscuring drama. History, and its near relative, Histrionics ... .[32]

This train of thought leads him to the observation – predictably questioned by his most rebellious student – that consequential historical events such as the French Revolution cannot really be said to have happened, their inaccessibility only masked – or perhaps replaced – by the construct of historical narrative: "Where then does the revolution lie? … . Does it lie in some impenetrable amalgam of countless individual circumstances too complex to be analysed? It's a curious thing … but the more you try to dissect events, the more you lose hold of what you took for granted in the first place – the more it seems it never actually occurred, but occurs, somehow, only in the imagination ..."[33] This claim is complemented by Crick's reflection on the ethics of such constructs and their inescapably political dimension. Speaking from the position of a representative of a family which has – purposefully or not – renounced any participation in great history,[34] Crick is unsurprisingly critical of the frenetic activity of history-making, stressing the inevitable violence involved in such

[31] Swift, *Waterland*, 232-3.

[32] Swift, *Waterland*, 34. Again, this observation on the official historiography concerned with great political events, is mirrored in private histories: "So there's no escaping it: even if we miss the grand repertoire of history, we yet imitate it in miniature and endorse, in miniature, its longing for presence, for feature, for purpose, for content." (34-5)

[33] Swift, *Waterland*, 121.

[34] See Swift, *Waterland*, 16: "For centuries the Cricks remain untouched by the wide world. No ambition lures them to the cities. No recruiting party or press-gang, foraging up the Ouse from Lynn, whisks them off to fight for King or Queen."

enterprises: "What every world-builder, what every revolutionary wants a monopoly in: Reality. Reality made plain. Reality with no nonsense. Reality cut down to size. Reality minus a few heads." He juxtaposes it with the "natural stuff" of curiosity, of "our love of life [which] is more anarchic, more subversive than any Tennis-Court Oath ever was."[35] Admittedly, however, at a crucial point of his narrative, Crick himself is lured by the deceptive strategy of subjugating experience to fantasy when he offers a neutralising explanation of his brother's suicide in front of his shocked father, unwilling to face its actual reason. The tragic deed is reduced to a statement for which Crick feels obliged to apologise:

> 'He's gone barmy.'
> (Forgive me, Dick. To malign your final gesture, your last recourse, with the taint of madness, to rob it of reality. I, if anyone, knew there was reason in your plight …).[36]

At the same time, the remarks denying any presence to events constituting the foundation of great and small histories, Crick's claims that non-eventfulness forms the traumatising core of the reality of the Fens, may themselves be reconsidered in interesting – and ethically productive – ways in the light of the non-existence of the real as postulated in Lacanian psychoanalysis. The real is, after all, what is not registered by the symbolic and, therefore, by definition, what cannot be said to be a positive entity in itself. This is not to say that the real is to be conceptualised as entirely unconnected to the symbolic, since its very incompatibility with the structures of that order situates the real as its limit, a point where the symbolic fails: "It is only because the symbolic cannot address the logic of trauma adequately that trauma is registered at all. While trauma itself may be proper to the real, the failure of its inscription is registered in the symbolic."[37] The category of the real may thus be applied to what is pushed to the margins of narratable reality by the discursive mechanisms shaping it, ensuring, indeed, the "monopoly in reality" that Crick scorns. This is not, however, to imply that the unregistered, the overlooked, the indescribable, the traumatic does not in fact take place, if only at the borderline of the paradigm of reality. After all, the narrator of *Waterland* is prepared to recognise the undeniable realness of incidents making up historical accounts or, indeed, the supposedly consoling symbolic

[35] Swift, *Waterland*, 178.
[36] Swift, *Waterland*, 304.
[37] Linda Belau, "Trauma and the Material Signifier," *Postmodern Culture* 11, no. 2 (2011), <http://pmc.iath.virginia.edu/text-only/issue.101/11.2belau.txt>, par. 32.

constructs themselves. The abstract structures created within the arbitrary limits of our social conventions do, after all, prove to be no less tangible than the real itself, as Crick testifies:

> And even as a schoolboy … it was still the fabulous aura of history that lured me, and I believed, perhaps like you, that history was a myth. Until a series of encounters with the Here and Now gave a sudden urgency to my studies. Until the Here and Now, gripping me by the arm, slapping my face and telling me to take a good look at the mess I was in, informed me that history was no invention but indeed existed – and I had become part of it.[38]

Therefore, as in most of Swift's prose, storytelling is contemplated in its dangerous potency for shaping our reality while the costs and limitations of the method are also exposed. Young Tom's hypocritical policy of "[a]ct innocent, you'll be innocent"[39] is – more than once – undermined by the horrifying events piercing through symbolic constructs. Perhaps the most prominent example is the situation which opens the main part of the narrative: the discovery of the body of Tom's friend, murdered by his half-brother Dick, floating in the river. After the corpse is pulled out of the water,

> because it is the recommended position for the resuscitation of the not-quite-drowned, Dad had him placed chest-down on the ground. And there … he began to press between Freddie's prominent shoulder-blades, to raise and lower his stiffened arms and to continue to do so for a full quarter of an hour. Not because he did not know, any less than Dick and I knew, that Freddie was dead, but because Dad, being superstitious, would never exclude the possibility of a miracle, and because this ritual pretence at resuscitation staved off the moment when we must face the indictment of truth.[40]

The application of the useless routine is an example of a symbolic institution used for protection against the truth the characters do not want to face, a way of shunning the shocking experience, a story they tell themselves and behind which they hope to hide. *Waterland* is, to a large extent, precisely an account of ceaseless intrusions of *tuché* into the order imposed by the symbolic upon the chaos of history. Storytelling is the only available means of attempting the resolution of this terrifying condition which, like the traumatic real in psychoanalysis, is an inaccessible kernel

---

[38] Swift, *Waterland*, 53.
[39] Swift, *Waterland*, 48.
[40] Swift, *Waterland*, 26.

around which language revolves, endlessly striving to contain it within its structures. Indeed, the whole tradition of storytelling in the Crick family is presented as being aimed at achieving a soothing effect in the face of the impossible encounter: faced with the unbearable, life-threatening trauma of the immediate experience of the meaningless world around them, the Cricks opt for storytelling as a means of coping with it. "To live in the Fens is to receive strong doses of reality. The great, flat monotony of reality; the wide, empty space of reality. Melancholia and self-murder are not unknown in the Fens. Heavy drinking, madness and sudden acts of violence are not uncommon. How do you surmount reality, children? ... How did the Cricks outwit reality? By telling stories."[41] In recounting memories from his childhood to his students, Tom Crick admits that "even in the no-nonsense and pragmatic twentieth century, this future schoolmaster quaked in his bed at night for fear of something – something vast and void – and had to be told stories and counter-stories to soothe his provoked imagination."[42] This memory in itself suggests the paradoxical presence of something threatening in what is described by the narrativising efforts of the family as "not there," but what still undermines this description. A monotonous, void reality filled with suicide, madness and acts of violence does, after all, appear to be something of a contradiction in terms.

At the same time, however, there are sections in the novel in which remaining (or perhaps stepping) beyond the domain of the symbolic is taken to correspond to preserving a lost paradise outside time. This state is distinctly associated with nature, since, as Crick puts it, "only animals live entirely in the Here and Now. Only nature knows neither memory nor history."[43] A basic contrast is postulated between the realm represented by the river Ouse, "a river which possessed as no man did, or still does, the secret capacity to move, yet remain" and "these two-legged intruders who by daring to transmute things into sound were unconsciously forging the phenomenon known as History."[44] This is another face of history contemplated in Crick's narrative, Natural History, "[w]hich doesn't go anywhere. Which cleaves to itself. Which perpetually travels back where it came from."[45] This idealising formulation echoes in a troubling way the design of Crick's grandfather, Ernest Atkinson, who "beats a headlong retreat, backwards, inwards, to Paradise, and starts to believe that only

[41] Swift, *Waterland*, 15.
[42] Swift, *Waterland*, 15.
[43] Swift, *Waterland*, 53.
[44] Swift, *Waterland*, 125.
[45] Swift, *Waterland*, 177.

from out of this beauty will come a Saviour of the World."[46] Atkinson's attempt to conquer time has disastrous consequences which affect even his rather distant descendants. Nevertheless, his gesture of renouncing the dominant ideology in favour of an ideal situated outside the discursive realm is, in a sense, repeated in Crick's own narrative, where the otherwise horrifying Here and Now is approached as a positive force counteracting the world-ordering, history-making efforts of the representatives of the symbolic, an anarchic tendency associated by Crick with "human nature" and curiosity. Crick detects in his audience the very quality of childlike curiosity which he glorifies as related to naturalness and innocence: "So you're curious. So you're curious. You'd skip the fall of kings for a little by-the-way scurrility."[47] Once again, the Natural History supersedes the constructs of culture. Crick's own youthful exploration of sexuality is situated within the context of tearing through the potentially dangerous illusions of the symbolic and the imaginary by means of the working of the same curiosity, which in his perception "[i]s a vital force. Curiosity, which bogs us down in arduous meditations and can lead to the writing of history books, will also, on occasion, as on that afternoon by the Hockwell Lode, reveal to us that which we seldom glimpse unscathed (for it appears more often – dead bodies, boat hooks – dressed in terror): the Here and Now."[48] As already signalled, this description of the Here and Now seems not only inconsistent with Crick's own presenting it as an unexpected, traumatising intrusion into the human construct of reality, which here becomes equated with a diametrically different image of the harmony and stasis of the idealised natural state. It also proves suspicious in the light of the narrator's ethical interests in relation to this point. His claims about nothingness, uneventfulness can hardly be seen as anything but part of an attempt to hide the "awkward" aspects of his reality. Crick's willingness to present a vision of nature as embodying the dream of a return to a pre-symbolic paradise goes against his distrust of analogous narratives in the political realm. In this, Crick, as a narrator performs exactly the kind of discursive violence that he claims to repudiate as a historian.

In his reflection on revolutionary tendencies recurring throughout political history, Crick remarks on their regressive character and points to their roots as lying in the human yearning "to return to that time before history claimed us, before things went wrong … How we pine for Paradise. For mother's milk. To draw back the curtain of events that has

[46] Swift, *Waterland*, 90.
[47] Swift, *Waterland*, 169.
[48] Swift, *Waterland*, 44.

fallen between us and the Golden Age."[49] From the Lacanian perspective, such a return is fundamentally impossible. The constitution of the speaking subject involves an irreversible loss and

> when being makes its appearance on the level of language, it must disappear under that language, it loses the reality of its being … The element lost in the process of becoming a human being is being itself, the pure being, the real, the thing without a name, leaving us with a basic lack as a condition for our becoming, which Lacan calls manque à être (want-to-be, or lack of being). Thus, right from the start, the subject is divided between the necessary loss of its being on the one hand and the ever alienating meaning in the Other on the other hand.[50]

The transgressive wish to return to a paradisiacal state of unity with the mother is illustrated by Crick's account of his brother's unsuccessful struggle to cope with the death of Helen Crick in a way which seems to imply Tom's embrace of the impossibility of turning back the flow of time. Dick is described as a "numbskull with the dull, vacant stare of a fish," who "[s]peaks half in baby-prattle, if he speaks at all. Never asks questions. Doesn't want to know. Forgets tomorrow what he's told today."[51] Deprived of the natural curiosity that both motivates the development of and subverts the achievements of civilisation, he is likened not only to an animal but also to his prized possession "a motor-bike, in its brainless efficiency, in its mechanical animation, bearing a pretty close resemblance to Dick himself."[52] One of the few issues that drag him out of this natural stasis is "that time his mother disappeared without explaining and never came back again … one day she took suddenly to her own bed and a little while later was seen no more. Now where could she have gone?"[53] Unable as he is to accept that "she's in that wooden box they lowered into the frost-hardened ground in Hockwell churchyard," Dick refuses to "believe she can have gone where she can't be retrieved. Perhaps she's hiding somewhere else. If they took her away in one box, perhaps she'll return in another … perhaps she's inside those bottles … Or perhaps what's inside the bottles will make her reappear ... ."[54] The bottles in question, left to Dick by his dying mother, contain an unusually potent beer brewed by his grandfather. Dick's experimentation with their content

---

[49] Swift, *Waterland*, 118.
[50] Verhaeghe, "On the Lacanian Subject," 176.
[51] Swift, *Waterland*, 209.
[52] Swift, *Waterland*, 210.
[53] Swift, *Waterland*, 210.
[54] Swift, *Waterland*, 210-1.

corresponds to the strategies undertaken by his father and his younger brother in managing the loss,[55] all three aiming to restore Helen, euphemistically referred to by Henry as "gone." Dick's strategy fails in the most spectacular fashion: the beer has a terrifyingly disorienting effect on the teetotaller, who

> cutting short his dance, stops hooting and cackling, sinks to his knees, puts a hand to his belly; feels his arms, his legs, his head to see if they are still there. His eyelids have never whirred so fast. A look of disbelief – of guilt, terror – crosses his face. A look not unlike the look he will give on a certain day by the Hockwell Lode, when something inside his woollen bathing trunks starts to stir unsuspectedly. He sits, but can't stay still, as if he'd never guessed quite what dangerous stuff he was made of, and he has to get away from it. But the only way to get away from it is to leap out of his own skin.[56]

Even without the analogy to Dick's unsettling realisation of his own sexual urges, the incident would be recognisable as an eruption of *jouissance*, a clear illustration of the dangers of trying to restore the unity with the mother. Dick's naïve understanding of substitution – as literally restoring the lost object – demonstrates the futility of any attempt to totalise the symbolic into a coherent order, ignoring its internal contradictions and incongruities which return inevitably in the form of the real.

More significantly, the dangerous stuff of which Dick is made reveals all the ethical and political investments of the novel's narratives. The magically destructive beer drunk in the incident is only a culmination of the disturbing influence of his grandfather Ernest Crick, who is also Dick's biological father. In the narrative, Ernest's Coronation Ale is a rather clear metaphor for his frustrated political ambitions. His pacifist warnings on the eve of World War I are disdainfully dismissed by the local community and his anti-imperial attitude infuriates the people unwilling to face the fact that a Golden Age is, indeed, coming to a close. The Coronation Ale, supposedly offered to the town in a gesture of reconciliation, in fact proves to have a profoundly upsetting effect on those who drink it, reducing the celebration of another event in the triumphalist narrative of unstoppable progress to utter chaos. Ernest is, it might be argued, "'real' in the strict Lacanian sense: a 'hitch,' an impediment which gives rise to ever-new symbolizations by means of which one endeavours to integrate and

[55] The former's repeated "rendezvous in the churchyard" (Swift, *Waterland*, 246) and the latter's attempts to replace her in the kitchen, termed "culinary necromancy." (Swift, *Waterland*, 247)
[56] Swift, *Waterland*, 249.

domesticate it … but which simultaneously condemns these endeavours to ultimate failure."[57] Pushed to the margins of the communal discourse, Ernest keeps returning, ceaselessly exposing its limits. The uncontrolled drunkenness induced by his beer literally dissolves social structures (turning against Atkinson himself when intoxicated fire-fighters fail to put out a fire at his brewery). Crick poses the rhetorical question of whether Ernest's anarchy-inducing beer did no more than to disclose the local community's own "inflammatory folly of their jingoistic ardour,"[58] but the community itself attributes the shocking goings-on to Ernest's revenge plot. A similar influence is then exerted by his daughter, Helen, whose sublime beauty disorganises a military parade. The latter event is neutralised in the accounts of its participants – including Helen herself – by being turned into a fairy tale and so is the incestuous relationship that Ernest has with Helen. Nevertheless, the disquieting effect of Ernest's transgressions is felt over the next generations: as a worshipper of beauty, Atkinson hopes to spawn the Saviour of the World, but his son refuses to close the gap in the symbolic. Indeed, Dick stands very much outside the order, himself real in the sense of causing interpretative efforts that are entirely beyond his comprehension. In his machine or animal-like nature as well as his uncertain grasp of language and inability to take in his mother's absence, he may be seen as a point where the symbolic fails. The stuff Dick is made of indeed proves dangerous when, thinking himself cuckolded by Freddie Parr, he murders the boy with a bottle of Coronation Ale. Thus, rather than offering an ultimate explanation, Dick in fact causes ever more Here and Now, enforcing the production of discourses which cannot but fail to capture his essence. Therefore, the obscene influence of Ernest Atkinson returns endlessly within and against the narrativising efforts of Tom Crick, whose story abounds in supernatural elements which emphasise the conventionality of the discourses enabling the repressions on which the delicate construct of social reality is founded. In fact, the inflammatory presence may be said to motivate the entire narrative, since Tom Crick is provoked into his attempt to voice all the unspeakable points of his family history when he is confronted with the ineffable fears of his student, Price, in whom he identifies a restlessness akin to that of his grandfather.

In this context, another level of heavy-handed irony is added to the question asked by a policeman investigating the kidnapping of the baby by

[57] Slavoj Žižek, "Introduction. The Spectre of Ideology," in *Mapping Ideology*, ed. Slavoj Žižek (London, New York: Verso, 1994), 22.
[58] Swift, *Waterland*, 272.

Mary Crick: "Look, sir, shall we go back to the beginning?"[59] Swift's characters are caught in a precarious balance between the headlong rush of events and their traumatising influence, arresting the passing of time altogether. Going back to the beginning is no more possible than the permanent drying of the Fenlands, which is why, near the conclusion of the novel, Tom Crick makes an analogy between the constant struggle to prevent water from overtaking the land and the notion of progress itself, which, according to him "doesn't progress. It doesn't go anywhere. Because as progress progresses the world can slip away. My humble model for progress is the reclamation of land. Which is repeatedly, never-endingly retrieving what is lost. A dogged and vigilant business. A dull yet valuable business. A hard, inglorious business."[60] The analogy may be extended to the functioning of the Lacanian subject, with the hydraulic metaphors (going back to Freud), and with the equally endless effort that it involves. As Dylan Evans observes in his discussion of *jouissance*, "it can be described as a kind of fluid with which the body is loaded at birth, some of which must be drained away in order to accomplish the 'work of civilization' (Freud) and allow entry into the symbolic (Lacan). This operation of drainage is what psychoanalysis designates by the term castration."[61] Even more significantly, this "drainage" is very much like that in which the inhabitants of the Fenlands are involved in that, as implied by Lacan's notion of surplus *jouissance*, "indicates that after castration has drained *jouissance* from the body, there is always a certain amount left over."[62] Ellie Ragland-Sullivan makes a similar observation, remarking on the paradoxical nature of loss in Lacan's theory: the functioning within social relations allows the restoration of some of the lost imaginary unity at the cost of serious concessions to the demands of the symbolic which is conditioned on maintaining the illusion of its coherence but by no means protecting the subject from the traumatising returns of the repressed real. [63] This is why, just as, in the words of Tom Crick, "you shouldn't go mistaking the reclamation of land for the building of empires,"[64] neither should it be presumed, in the light of his

---

[59] Swift, *Waterland*, 272.
[60] Swift, *Waterland*, 291.
[61] Dylan Evans, "From Kantian Ethics to Mystical Experience: An Exploration of Jouissance," in *Key Concepts*, ed. Dany Nobus (New York: Other Press, 1999), 11.
[62] Evans, 12.
[63] Ellie Ragland-Sullivan, *The Logic of Sexuation. From Aristotle to Lacan* (Albany: State University of New York Press, 2004), 26.
[64] Swift, *Waterland*, 291.

narrative, that language can ever dispel the trauma of leaving the supposed pre-symbolic paradise.

References

Belau, L., 2011, "Trauma and the Material Signifier," *Postmodern Culture* 11, no.2 (2011), <http://pmc.iath.virginia.edu/text-only/issue.101/11.2belau.txt>.

Evans, D., 1996, *An Introductory Dictionary of Lacanian Psychoanalysis*, London and New York: Routledge.

—. 1999, "From Kantian Ethics to Mystical Experience: An Exploration of Jouissance," in Nobus, D., ed., *Key Concepts of Lacanian Psychoanalysis*, New York: Other Press, 1-28.

Fink, B., 1995, *The Lacanian Subject. Between Language and Jouissance*, Princeton: Princeton University Press.

Freud, S., 1963, "The Introductory Lectures on Psycho-Analysis," in Strachey, J. ed., and trans., *The Standard Edition of the Complete Psychological Works of Sigmund Freud*, London: Hogarth Press and Institute of Psychoanalysis, vol. 16.

Homer, S. 2005, *Jacques Lacan*, London: Routledge.

Hutcheon, L., 1988, *A Poetics of Postmodernism: History, Theory, Fiction*, London & New York: Routledge.

Lacan, J., 1981, *The Language of the Self: The Function of Language in Psychoanalysis*, Anthony Wilden, trans., Baltimore, Maryland: John Hopkins University Press.

Ragland-Sullivan, E., 2004, *The Logic of Sexuation. From Aristotle to Lacan*, Albany: State University of New York Press.

Sharpe, M., 2010, "Lacan, Jacques," *Internet Encyclopedia of Philosophy*, <http://www.iep.utm.edu/lacweb/>.

Swift, G., 1983, *Waterland*, London: Picador.

Verhaeghe, P., 1988, "Causation and Destitution of a Pre-ontological Non-entity: On the Lacanian Subject," in Nobus, D., ed., *Key Concepts of Lacanian Psychoanalysis*, New York: State University of New York Press, 164-189.

Wright, E., 2000, *Lacan and Postfeminism*, Cambridge: Icon Books.

Žižek, S., 1994, "Introduction. The Spectre of Ideology," in *Mapping Ideology*, London, New York: Verso, 1-33.

CHAPTER SEVEN

ON TIME MANIPULATIONS AND THE MEANING OF THE REAL: THE CASES OF ROBERT GRAVES, MALCOLM ROSS AND ALEJO CARPENTIER

ALICJA BEMBEN

> "[Philosophizing] is 'nothing but' the transformation of the same few simple questions. But those who wish to transform must bear within themselves the power of a fidelity that knows how to preserve. And one cannot feel this power growing within unless one is up in wonder. And no one can be caught up in wonder without travelling to the outermost limits of the possible. But no one will ever become the friend of the possible without remaining open to dialogue with the powers that operate in the whole of human existence."[1]

David Daiches' *The English Novel and the Modern World* opens with the claim that the changes that took place in the socio-economic life of Europe in the first half of the twentieth century impacted the shape of the English novel of this period. "[T]he breakdown of public agreement about what is significant in experience and therefore about what the novelist ought to select, the new view of time, and the new view of the nature of consciousness" triggered, among others, the introduction of new narrative techniques and new points of view, as well as reconstitution of the relationship between the author and his subject.[2] Such practices – and,

---

[1] Martin Heidegger, "For Edmund Husserl on His Seventieth Birthday, April 8, 1929," ed. and trans. Thomas Sheehan, <http://www.clas.ufl.edu/users/burt/pdfs/e-8-iv.pdf>, 3.

[2] David Daiches, *The English Novel and the Modern World* (Chicago, London: The University of Chicago Press, 1965), 1.

I would argue, practices going far beyond the English novel of the first half of the twentieth century – might be considered as manifestations of the writers' endeavours to deal with the shifts in thinking about the real by means of fiction. The wearing out of the hitherto widely-acknowledged concepts of what is real, the collapse of the thus far contested axioms and the sharpening of an overwhelming sense of reality losing its meaning in the face of the seemingly void future made writers of the age transform the novel into a field of experiments where we can observe them reworking the basic concepts, notions and categories on which their perceptions of the real would be built.

Among these *time* seems to be especially important. Recognizing this significance, this essay focuses on time-constructs which shape the narratives of Robert Graves's *Seven Days in New Crete,* Malcolm Ross's *The Man Who Lived Backward* and Alejo Carpentier's *The Lost Steps.* My purpose is, first, to "flesh out" the time experiences these texts offer and then to juxtapose them with the time-concepts of Mircea Eliade, Reinhardt Koselleck and John Bigelow[3] in order to make the philosophical linings of the time-concepts devised by Graves, Ross and Carpentier visible. Such an analysis serves to substantiate the thesis that the exhaustion of the potential of the clock time conventions to imbue events with meaning could be interpreted as a stimulus for these writers to move away from *mechanical time* towards *natural time*. Because the latter seems to retain the potential not only to generate, but also to restore lost meaning, then, consequently, the manipulations of clock time carried out in the texts of Robert Graves, Malcolm Ross and Alejo Carpentier might be read as discursive acts that aim at ensuring the meaningfulness of the real.

The point of departure for the following analysis is Heidegger's claim about the dependency between time and making sense of the real. "Time is primordially the horizon of the understanding of Being,"[4] writes the philosopher, and pushes his readers to recognize that the way one conceptualizes time provides the framework for one's understanding of reality. Applying this perspective to the changes mentioned by David Daiches, makes it possible to think about the pre-war clock-time concept in terms of the horizon that failed to enable seeing the real as meaningful. World War I was not comprehensible in terms of a uniform and linear clock time with its neat division into the past, the present and the future. Regardless of the explanations provided, the post-war world was permeated with a sense of meaninglessness. For the real to become meaningful once

[3] Arguments supporting the above are provided in the further parts of this essay.
[4] Martin Heidegger, *Being and Time* (New York: Harper & Row, 2008), 18.

again, it was necessary to devise such concepts of time that would frame the post-war chaos into a meaningful entirety. The writers, whose works are discussed in this study, used fiction to elaborate their own concepts of time, concepts which would allow them to see the real again as meaningful. They turned their backs on clock time and embarked on intellectual journeys to natural/mythical time,[5] which were constantly powered by an ever-reviving nature, and seem to retain the potential to generate as well as to restore the lost meanings.

Robert Graves's *Seven Days in New Crete* is narrated by Edward Venn-Thomas – a twentieth-century poet who wakes up in a remote future,[6] in the land of New Crete. When travelling across the realm, he embarks upon numerous adventures, falls in love with a witch, gets involved in political machinations, and, amidst all this, gains knowledge of the quaint custom that governs the world he discovers. The custom in question is a set of rules formulated by poets and magicians inspired by their Mother-goddess.[7] Following these rules, the New Cretans live in kingdoms, each further divided into five social groups – the captains, the recorders, the commons, the servants and the magicians.[8] Importantly, however, within the kingdoms of the New Crete the power is held by queens who appoint new kings for themselves upon each half of the moon cycle.[9] Edward discovers the intricacies of the custom when on the seventh day of his sojourn in New Crete, towards the end of the moon cycle, the Mother-goddess reveals herself to him, engages him in a conversation on the fate of her realm, makes him destroy it and return to twentieth-century Europe. During his stay in the future, the narrator is fascinated by the new place and rapidly absorbs knowledge about it. Yet, despite his fascination, a sense of dissonance always accompanies him. Summoned from the age of mechanical time, he finds himself among people who do not recognize it: to survive, he must adapt to the new environment, which requires him leaving behind clock time and embracing the experience of temporality offered by the New Cretan world.

---

[5] Replacing clock time with natural time is only one of the tendencies among post-war writers, though, perhaps, one of the most interesting ones.

[6] Robert Graves, *Seven Days in New Crete* (London: Quartet Books, 1975), 40.

[7] Graves, *Seven Days in New Crete*, 18: "Custom is the governing principle …" and 19 "Custom here is based not on code of laws, but for the most part on the inspired utterances of poets; that is to say, it's dictated by the Muse, who is the Goddess."

[8] Graves, *Seven Days in New Crete*, 16-17.

[9] Graves, *Seven Days in New Crete*, 98-99.

Thinking about time in his own era, Venn-Thomas recalls the hastiness of his contemporaries, stresses their constant complaints about lack of time and their blithe disregard for the quality of the present, as well as their excessive preoccupation with what has already happened and what is still to come.[10] Such an approach stands in sharp contrast with what he observes in the New Crete. Firstly, Edward notes that unless an event is of direct and crucial importance to the current situation, the New Cretans have no regard for it. They ignore it, recording only those past events whose knowledge is essential to their survival, or those they consider to have been inspired by the goddess.[11] This insouciance – initially incomprehensible to the narrator – turns out to be the result of the White Goddess' abolishing clock time:

> As we walked home, I asked See-a-Bird: 'What year are we in?'
> 'The year before lap year.'
> 'Yes, but what's the date?'
> The Interpreter intervened. He explained to See-a-Bird that in my age we counted the years publicly and celebrated every first of January with a post-mortem on the Old Year speculations on the New.
> 'Here we have no public date,' See-a-Bird told me. 'The chief recorder keeps a count of years in the archives, but it isn't published and nobody but he and his assistants could calculate how many years have elapsed since the foundation of New Crete. We also consider it highly improper to mention anyone's age or to count the number of years that he has held office or been married. In the same way we make no record of hours and minutes, as I believe you do with clocks and watches. We observe the phases of the moon; we distinguish morning from afternoon and afternoon from evening; we keep the days of the week; we mark the passage of the seasons; and the two parts of our double year end with the first full moon after the longest day and the first full moon after the shortest day. But time in an absolute sense was abolished on the same occasion on which it was agreed to abolish money; for the poet Vives pleaded passionately:
>
> *Since Time is money,*
> *Time must be destroyed:*
> *His sickle and hourglass*
> *Are in pawn to evil.*
> *Nimuë, save us with your bow again.*

[10] Graves, *Seven Days in New Crete*, 9, 50, 80-81, 85-86.
[11] Graves, *Seven Days in New Crete*, 18.

> 'Then at what time do children go to school in the morning?'
> 'When the bell rings.'
> 'And when does it ring?'
> 'When the first three children have arrived.'
> 'And how long do you boil an egg?'
> 'Until the sand runs out of the egg glass.'[12]

To put it simply, in the New Cretan world, mechanical time is ignored. Their goddess – an omnipotent and all-wise law-giver[13] and the source of the custom which the Cretan society adopted – "designed" the New Cretan lack of sense of mechanical time and replaced it with natural/mythical time. Edward learns that in New Create it is conceived in terms of an incessant repetition of the Goddess' custom. According to it, at the beginning of each lunar year the queen that is an embodiment of their Goddess selects a king who is to accompany her for the first half of the lunar year. This is a period of growth, when all social activities aim at producing and developing whatever is necessary for the proper functioning of the society. But when the second part of the lunar year is to commence the thitherto king is replaced by another one who is to accompany the queen for the remaining time. This period is a time of decline when what the society has earned is spent, used up or consumed. Finally, each lunar year ends with a celebration honouring the Goddess.[14]

On the one hand, following the custom, his new companions treat each beginning of a lunar year as yet another beginning of their world. On the other, the White Goddess celebration performed at the end of each year brings their current world to an end and thus, through purification from what has grown old, it enables and simultaneously commences the beginning of a new one. Moreover, although the New Cretans do not measure time, ignore it and seem to live in the eternal present, they have a set of customs that regulate the timing of virtually every aspect of their lives[15] – customs which are by no means dependent on clock time, but which are to reflect the workings of nature. They do recognize lunar months and years, but the coming and passing of these do not push the society to count or measure time: they live by the rhythm of the signs from the Goddess, who tells her disciples how and when to follow her orders.

---

[12] Graves, *Seven Days in New Crete*, 63.

[13] Graves, *Seven Days in New Crete*, 278.

[14] The custom presented in *Seven Days in New Crete* is at the same time a description of the White Goddess Myth.

[15] See the quote above – even coming to school is organized according to the custom and not according to clock time.

Thus, the New Cretans perceive events as a part of the natural order instilled by the Goddess and Edward, plunged into the natural/mythical time of New Crete, soon shares their experience.

In order to expose the philosophy underlying the Gravesian time construct, I would like to juxtapose Graves's idea of the White Goddess Myth, on the basis of which his book is designed, with an idea bearing a marked similarity: Mircea Eliade's concept of *illud tempus*.[16] Mircea Eliade's understanding of the beginning of the beginnings of humanity is based on the assumption that

> the chief moments indubitably remain the purification through the scapegoat and the repetition of the cosmogonic act by Yahweh; all the rest is only the application, on different planes answering to different needs, of the same archetypal gesture: the regeneration of the world and life through repetition of the cosmogony.[17]

His claim is then that ancient cultures crafted for themselves descriptions of their own cosmogonic acts to see the surrounding world as *ordered*. Although such creation myths would vary in details, they would, nonetheless, share a general frame. According to Eliade, the motif linking them is the recurrent motif of self-sacrifice of an individual effecting the re-enactment of the cosmogonic act. Time for these peoples was then a series of re-enactments of their cosmogonic act, that is of *illud tempus*.

Graves's understanding of cosmogony fits Eliade's theory not only at first sight – the concepts take the cycle of life and death as their most general framework, point to the creator of the world renewing it and mention the requirement of a sacrifice to achieve the renewal – but the most important link between these two visions is that they both emphasize the mythical character of the time of creation. On the basis of these similarities and Eliade's extension of his theory to include the notion of the *illud tempus* – mythical primordial time[18] – it is possible to describe the time-construct that Graves offers in the novel by means of an analogy to Eliade's primordial time.

If, as Eliade maintains, ancient cosmogonies are descriptions of *illud tempus* and Graves's description in *Seven Days in New Crete* is a variation on his idea of cosmogony, then consequently, it might be argued that the time-construct in the novel is a variation of *illud tempus*. Obviously, it

---

[16] Mircea Eliade, *Cosmos and History: The Myth of the Eternal Return*, trans. Willard E. Trask (New York: Harper and Brothers, 1959), 73.
[17] Eliade, *Cosmos and History: The Myth of the Eternal Return*, 62.
[18] Eliade, *Cosmos and History: The Myth of the Eternal Return*, 73.

may also be argued that due to the fact that the narrator comes from twentieth century Europe and thus from the age of clock time, that might not be the case, but a closer look at Graves's theories allows one to dispel the doubt. Firstly, we are instructed that the Late Christian era is supposed to be understood in terms of mechanical time – talking about his epoch, Venn-Thomas concludes that "[i]t's the mechanization of life that makes our age what it is: science and money combine to turn the *wheels* round faster and faster."[19] Secondly, we learn that Edward's seven-day stay in New Crete is a part of one of the cycles by which the realm functions. This might bring us to the conclusion that the overall picture of time offered in the novel is a combination of mechanical time (Europe) and mythical time (New Crete). However, a reference to Graves's White Goddess Myth allows one to rectify this surmise. The Late Christian era discussed in the novel is a repetition of the *illud tempus* in the same way in which the reign of New Cretan queens and kings is. Just as the time of growth is followed by the time of decay in New Crete, so Europe of the post-war time is in the decadent phase of the same cycle of life and death, in which mythical time degenerated into clock time.[20] It might be inferred, then, that Graves's understanding of time renders history as a series of variations on the White Goddess Myth. If this is the case, then clock time is a concept which Graves equates with the phase of degeneration and the loss of the potential to create meanings. Consequently, its "abolition" and return to mythical time denotes, on the one hand, coming back to the time of growth, when the potential of the force of nature to (re)regenerate and (re)create is on the increase. On the other hand, it is also a time when nature's growth parallels the growth of the potential to create new meanings.

Inasmuch as Graves is an interesting case of a writer who recasts time from a mechanical to a natural/mythical one in order to see the real as meaningful, the case of Malcolm Ross presents yet another way of time manipulation to achieve the same aim. The narrator of the novel *The Man Who Lived Backward* lives forward for himself, but backward for other people. Born in French Montmédy in June 1940 "at the darkest hour of World War II," Mark Selby lives for seventy-five exceptional years and ends his life in an attempt to prevent President Lincoln's death on 15 April 1865. We learn about the peculiarities of the last forty years of his existence – and also, sketchily, about his youth – from a diary which he

[19] Graves, *Seven Days in New Crete*, 10. My emphasis.

[20] Edward Venn-Thomas might appear to be a spur of the mechanical-time origin for closing one cycle in Europe and one in New Crete.

starts on 17th February 1901 and finishes on 16 April 1865.[21] The extraordinariness of Selby's life consists in that although he wakes up at dawn, goes through the day and then falls asleep, his next morning is not that of a consecutive day, but of the day before. Thus, even though, in his own eyes, he ages with the passing years, for those who surround him he gets younger and younger every day. In 1934 he is a child of six, supported by an American gentleman – his friend and will executor – and nursed by Helen Selby, a mother-like figure and his aged wife.[22] During 1923-1903 the couple is separated for "that terrible transition period between the relationship of man and wife and that of mother and son."[23] This time Selby spends travelling, celebrating the victory of the Great War, then fighting in it and making money. In 1903, at the age of thirty-seven, he marries Helen Smith whom he meets in 1901.[24] From that year up to 1865 the diary gives detailed accounts of Mark's encounters with a variety of people – actresses, journalists, Woodrow Wilson, religious leaders, Walt Whitman, little Helen, one of his French ancestors and, finally, with an ex-soldier.[25] From Selby's perspective, his life is "normal," as he grows older and older, and yet he remains acutely aware that, in fact, it is he who lives the reverse order of days and seems to grow younger and younger for his wife and friends.[26] His life is interwoven with lives of other people whose other flow of time he recognizes and adjusts to. Extremely covert about his condition, the narrator reveals his secret to his wife/mother figure Helen and to the ex-soldier, his will executor.[27]

Of Helen we know that at the age of sixteen she meets and falls in love with the thirty-nine-year-old Mark Selby. After her initial disbelief in the narrator's condition is dissolved by Selby's future-forecasting and getting younger and younger she becomes his wife,[28] helps him with his investigation into the causes of the reverse life and stands by his side even when he returns to the womb of his mother.[29] Helen's role in Mark's

[21] Malcolm Ross, *The Man Who Lived Backward* (New York: Farrar, Straus and Company, 1950), 454-460.
[22] Ross, *The Man Who Lived Backward*, 15-23.
[23] Ross, *The Man Who Lived Backward*, 342.
[24] Ross, *The Man Who Lived Backward*, 19-21.
[25] Ross, *The Man Who Lived Backward*, 40-230.
[26] Ross, *The Man Who Lived Backward*, 16, 27.
[27] Ross, *The Man Who Lived Backward*, 354. Also Walt Whitman knows the secret – the poet appears in the story episodically.
[28] Ross, *The Man Who Lived Backward*, 20-21.
[29] Ross, *The Man Who Lived Backward*, 459. The plot of the novel is very convoluted – we do not get a neat from-A-to-Z story, but rather a multiple-perspective and patchy one.

coming to terms with his time experience is twofold. On the one hand, it is thanks to the upbringing she has given him that the narrator perceives his peculiar life as relevantly manageable. Selby's abilities to rationalize the backward order of days are the results of Helen's exertions to show him that two different and yet simultaneous flows of his time, i.e. day-to-day backward living and within-day forward living, may be reconcilable. A mixture of acceptance and ignorance brought by her love gives Mark a sense of appeasement: "out of the moments of crisis and Helen's comfort there comes to me now a feeling of continuity which I never had before. Against time though I travel, there will be no variations in the schedule." Helen is also the one who helps the narrator to deal with the third flow of time he experiences, that is the forward flow of time of other people. Her relationship with Selby teaches him to incorporate into his own temporal perspective the time of those people who affect his life. Mark does that by sharing emotions with his companions. He seeks "an authentic meeting of minds"[30] for he believes that "the common denominator of the emotions in man is what counts."[31] That is why he is able to understand and be with Helen:

> I am no spectre. I can't relieve my past. Whatever was real in it – palpable flesh and felt breath – lies in time's future where I may not enter. But I carry that future in my mind, and the mind is a reality. A unique one. No one owns my mind but I. There is identity in it with people, events, emotions which I share with billions of others because they are related only to me at a specific moment of time and place. [32]

This focus on specific moments of time helps Selby to understand and incorporate other people's flow of time into his own time experience. Although he does not share the forward succession of days with other people, he can share certain moments with them.

However, Helen is not the only person to know Selby's secret. Just a few days before his death in April 1865, Mark meets an ex-soldier to whom he reveals his secret. The relationship between the two men develops for their flows of time are possible to converge: "[t]his moment which you and I are sharing is uniquely itself for both of us ... there is a response between us,"[33] explains Mark. It is upon such moments when

[30] Ross, *The Man Who Lived Backward*, 360-361.
[31] Ross, *The Man Who Lived Backward*, 187.
[32] Ross, *The Man Who Lived Backward*, 340.
[33] Ross, *The Man Who Lived Backward*, 24.

flows of time come together, that Selby builds and maintains his relationship with the ex-soldier.

Finally, but perhaps most importantly, Selby sees one more flow of time and that is natural time devised by God. Regardless of whether he lives forward for himself and backward for other people, and of whether other people live the other way round for him, they all are according Selby embedded in the "pool of time"[34] created by God. Reaching this thought, finally, the narrator manages to reconcile with his peculiar condition and conclude: "I shall live out my days until I am old and ready for the reward of death, as other men, in sequence."[35] The differences in the individual flows of time which make Selby think of the inadequacy, purposelessness and meaningfulness of his life disappear in God's pool of time where all people function in one universal and natural pattern – that is the cycle of life and death.

If Ross offers in his text an experience of time that is based on flows of time, then to delineate his concept of time it might be useful to juxtapose the experience he offers with a concept of time that is formulated around the very same idea. Using Reinhart Koselleck's concept of "layers of time,"[36] I would like to argue that the time construct offered in *The Man Who Lived Backward* is that of converging flows which derive from natural time. According to Koselleck's theory, humans exist in a pregiven natural time and these are "nature-bound, traditional rhythms of time," such as days and seasons, which are the prerequisites for "human history and its historiography." It is that, Koselleck writes, we borrow from natural time to construct our layers of historical time

> Historical time, if the term is to have a meaning, is tied to social and political units of action, to particular acting and suffering human being, and to their institutions and organizations. They all have certain inherent modes of performance, each of which has its own temporal rhythm … . We might speak, not of one historical time, but of many that overlie one another.[37]

---

[34] Ross, *The Man Who Lived Backward*, 360-361.
[35] Ross, *The Man Who Lived Backward*, 224-226.
[36] Mishkova Diana, "Scale and Cognition in Historical Constructions of Space," in *Historein* vol. 10 (2012), <http://www.nnet.gr/historein/historeinfiles/histvolumes/hist10/historein10-mishkova.pdf>, 3.
[37] Reinhart Koselleck, *Practice of Conceptual History. Timing History, Spacing Concepts*, trans. Todd Samuel Presner and Others (Stanford: Stanford University Press, 2002), 102-110. See also page 106: "Because history itself remains embedded in time periods that are pregiven by nature, historiography is likewise unable to dispense with them."

Should we take into consideration the time flows that Ross describes, then the already-mentioned fourth type of the flow of time, i.e. "pool of time" might be interpreted in terms of Koselleck's idea of pregiven natural time. Not devised by humans, it is the God-given foundation for humans to create within it their individual senses of the flow of time. Consequently, time flows – parallel to Koselleckian layers of "historical" time – that derive from it and might be discerned in the story are Selby's reversed order of days and the forward flow of time within days as well as the forward succession of time of other people. Although each of these flows/layers has its own temporal rhythm, they all are ultimately framed by natural time. And if Selby's disappointment with life stems from his inability to generate meanings that would make sense in all his flows of historical time, then it is his final reversing to the all-encompassing natural time that enables him to make sense of his life. Despite the internal maelstroms in its historical aftermaths, the life and death cycle seems for Ross to be an unchanging and universal point of departure for the creation of meanings.

Nonetheless, the potential to generate meaning is only one of the qualities that natural time seems to have. Yet another one is that it may also enable the restoration of the meaning that has been lost, and Alejo Carpentier's manipulation of the flow of time in *The Lost Steps*[38] seems to aim at that very idea. The writer acquaints us with a concept of time that, by flowing and ebbing, expands the present and allows for the retrieval of the meanings once lost.

The narrator of the novel – a disillusioned and unfulfilled musicologist – believes himself to be trapped between the cogs of his daily routine. Commercial music-making, cheating on his wife, as well as attending her performances and parties lock him "among clocks, chronographs and metronomes."[39] This imprisonment lasts until he meets an old teacher – the Curator – who wishes to send him on an expedition to the South American jungle for "unique idiophones." The meeting pushes the already hyper-acutely time-aware narrator to the verge of mental exhaustion. It is so, because on the one hand, the Curator embodies the "generation which saw things in terms of 'the sublime,'" and thus painfully reminds his former student of the shallowness of his present life. On the other, the Curator's inquiry about the narrator's theory on the origins of music

---

[38] Alejo Carpentier, *The Lost Steps*, trans. Harriet de Onis, Minerwa (London: Minerva, 1991). The original text is written in Spanish, I have consulted the English translation by Harriet de Onis.
[39] Carpentier, *The Lost Steps*, 10.

revives the musician's memories of his lost, yet blissful youth.[40] Given an opportunity to restore meaning to his life, the musicologist takes advantage of the offer and runs away into the jungle. Leaving civilization behind, he travels, both into the heart of the jungle and through all the epochs until he reaches "the devil's cauldron"[41] – a crater in which plants that saw the birth of the earth still grow, and which he sees as the source of time.

The whole story the traveller recounts starts with his experiencing time in which the past, present and future are seen as separate. He feels trapped in the present moment that appears to him as if built of a series of meaningless and repetitive actions. It is neither linked with the past – which is lost and irretrievable, and no meaning can be derived from it – nor with the future – being stuck in the recurrent present moment means there is no other future than the ever-repeating present moment. From the narrator's account, we learn that it is a period of one week. During the weekdays he is caught in the automatism of his work and races between various things-to-be-done-immediately. Free evenings and weekends are reserved for his wife who, as an actress, requires her husband to accompany her to numerous social events. There is also "the Seventh Day": on Sundays, the couple fulfil their marital duties – and although they are executed out of habit, urge or pangs of conscience, the intercourse evokes in the narrator memories of both lost intimacy and of his *dear* wife.[42] The present, restarted each Sunday, is the prison from which there is no escape. It offers constant torments by means of either lurks of the irretrievable yet imbued-with-meaning past or looms of the future that is drained of any meaning.

Although the narrator perceives the past as lost and no longer existent, its remnants are hidden in objects (musical instruments, the text of *Prometheus Unbound*, university exhibitions), events and people (the encounter with the Curator, women's gestures and orchestra players).[43] These, along with the pressing sense of being trapped, lend the musicologist an impetus to set off on a journey which alters the way he experiences time. On the one hand, from his remarks we learn that the expedition starts on 7 June and ends before 18 July.[44] Therefore, the narrator is perfectly aware of the forward flow of time throughout the

[40] Carpentier, *The Lost Steps*, 21-32.
[41] Carpentier, *The Lost Steps*, 205. The narrator imagines he travels into the past – that each next place he visits is from a more and more distant epoch.
[42] Carpentier, *The Lost Steps*, 6-10.
[43] Carpentier, *The Lost Steps*, 8-14, 16-35.
[44] Carpentier, *The Lost Steps*, 37, 239.

whole adventure. The people he meets, the places he visits, the events he takes part in are described chronologically in his journal. At the same time, having left his capital city,[45] Los Altos, La Hoya, Santiago de los Aquinaldos, Puerto Anunciación and El Pintado behind, to finally reach Santa Monica de los Venados, he claims to have "traveled through ages" to places located in more and more distant epochs.[46] Thanks to the journey, the narrator's present starts changing. Firstly, it ceases to be the seven-day present and becomes a sort of *the under-construction-present*. Since he feels no longer constrained by mechanical time and can only count the passing days, the narrator builds his sense of the present of the oncoming days – days which both the future and the past bring simultaneously. Having reached his old hometown, the narrator reacquaints himself with the language he knew in his childhood. On the one hand, each day of this journey is day from the future that contributes to his present. On the other hand, it is also a journey back into his childhood. Each day he moves back into the past is also one day that is not lost from his present but retained in it. Childhood, however, is not where the narrator stops. As if in the domino effect, he moves to even more remote times – constantly gathering the days that the future brings and retaining the days which should become his past.

When he reaches the country where people "are accustomed to living with Rousseau and the Inquisition, with the Immaculate conception and *Das Kapital*"[47] and whose present is not made up of only what exists physically right now, but is simultaneously composed of what chronologically existed in different periods of time, the effect on the narrator's own experience of time is that his present expands even further, beyond his own life:

> Up to that moment the change from the capital to Los Altos had been for me a kind of return, through this renewed experience of ways of life, flavours, words, things, which had left a deeper brand on me than I would have believed, to the years of my childhood, to my adolescence and its first awareness ... . But now I was moving beyond the images that had met my eyes at a time when I ceased to know the world only through the sense of touch ...[48]

From mere recollections and reflections on the onetime architectural styles, the narrator moves on to interpreting events and happenings around

---

[45] The text does not give the name of the capital.
[46] Carpentier, *The Lost Steps*, 277.
[47] Carpentier, *The Lost Steps*, 50-51.
[48] Carpentier, *The Lost Steps*, 77.

him as if they were embodiments, or equivalents, of what had already happened a long time ago. For instance, the revolution that he has entangled himself into turns out for him to be

> more akin to a religious war. Because of an incredible chronological discrepancy of ideals, the conflict between the conservatives and those who seemed to represent extremist tendencies gave me the impression of a kind of battle between people living in different centuries.

Although the revolution "enters" the narrator's present from the future's side, it is concurrently interpreted as an element of a distant past. The musicologist incorporates into his present a fragment of the future, i.e. the oncoming revolution, but also a piece of a quite remote past, i.e. the long-forgotten revolution. As the journey continues, the narrator keeps adding to his present ensuing moments from both, the future and the past. For example, when around the feast of Corpus Christi he, his lover Mouche and their fellow traveller Rosario arrive at a town[49] they meet waves of people who sweep across the place. First, they encounter a group of prostitutes who seem to be "a mixture of a carnival wench and a St. Mary of Egypt without odor of sancticity." One of their clients is Yannes - a Greek taken as if from Homer's *Odyssey*. Secondly, the narrator starts thinking about Rosario in terms of an ancient saint like St. Cecilia or St. Lucy, whereas Mouche becomes an unwanted burden of modernity. Thirdly, the reading of a book by Fray Servando de Castillejos takes him three centuries back into the epoch of missionaries investigating the Orinoco River.[50]

Should these events be put in a chronological order, we may observe that with each successive happening, the narrator moves also further and further back into the past. Time flows forward when the narrator meets the prostitutes but simultaneously ebbs much further into the past when they are interpreted as carnival wenches. Time flows forward once again when Yannes appears, but only to ebb to the times of Homer. The whole journey is a repetitive swaying of time until the narrator reaches "the devil's cauldron." On seeing it, he finally renounces "the measured time" and accepts the eternal present.

> Now Fray Pedro led me to the other end of the Signs and pointed out on that side of the mountain a kind of crater in whose depths horrendous

[49] Carpentier does not offer names for many places he describes. The capital city he mentions and the above mentioned town remain nameless as well.
[50] Carpentier, *The Lost Steps*, 101-110.

> plants proliferated ... ." These are the plants which have fled from man since the beginning," the friar told me, "the rebel plants, those which refused to serve him as food, which crossed rivers, scaled mountains, leaped the deserts for thousands and thousands of years, to hide here, in the last redoubts of Prehistory."
>
> In silent amazement I gazed on what in other places was fossil impression or slept in the petrification of coal, but here continued alive in a timeless spring antedating the days of man, whose season might not be that of the solar year. Perhaps their seeds germinated in hours; perhaps they took half a century. "This diabolical vegetation that surrounded the Garden of Eden before the Fall." Leaning over the devil's caldron, I felt the vertigo of space. I knew that if I let myself come under the spell of what I was looking upon here, this prenatal world, I would end up by hurling myself down, burying myself in this fearful density of leaves which would one day disappear from the planet without having been given a name, without having been re-created by the World. They might be the creation of gods that came before our gods, gods on trial, clumsy in their works, unknown because they were never named, because they never took on shape in the mouth of man.[51]

Thus, the overall experience of time offered in *The Lost Steps* is that of an expanding present. Of great importance for this argument is also the question of what construct of time lies behind such an experience. In order to answer it, I would like to refer to John Bigelow's idea of "presentism."

According to Bigelow, "nothing exists which is not present,"[52] and there exists only "a moving Now" – an individual present moment moving through a succession of presents which includes parts of the future becoming the present and leaves behind fragments of the present which become the past. Juxtaposing the experience of time presented in Carpentier's text with this idea allows one to deduce that mechanical time in which the narrator is trapped might be interpreted as a small cage of the "moving Now" that leaves the past behind and anticipates a future no different from the present, that leaves certain meanings behind and expects no new meanings to come, that permits futile life but disallows building meanings. The creation of a new extended and nature-based Now might be thus interpreted as a discursive act, on the one hand, of retrieving or bringing into existence once again the meanings that were lost and on the other, of admitting the possibility of creating new meanings. Altogether, if – as it might be inferred from Carpentier's construction of time experience

[51] Carpentier, *The Lost Steps*, 205-206.

[52] John Bigelow, "Presentism and Properties," in *Noûs* vol. 30, Supplement: Philosophical Perspectives, 10, Metaphysics (Cambridge and Oxford: Blackwell Publishing, 1996), 35-52.

– meanings cannot be created in the present which is but a clock-time cage, then their generation requires such a concept of time that would be more than a small prison. Thus, it seems that according to the author of *The Lost Steps*, for reality to be meaningful it is necessary to abandon the concept of clock time. In its place the author offers the concept of the eternal present, in which temporal rhythms derive from nature, in which no meaning is lost and in which the potential to create meanings can be continually exploited.

In each of these three texts, time manipulations aim at dealing with the clock-time-construct that does not enable any more to imbue events with meaning and offering alternative constructs that might replace it. Firstly, the past which initially emerges from these texts seems to be either lost or meaningless. Renouncing clock time initiates the change in "the horizon of understanding" and adopting natural/mythical time enables one to (re)create or revive meanings which have been lost. Secondly, although the present appears as if drained of meaning from the perspective of mechanical time, from the angle of natural time it turns out to be the point of departure from which the writers start rebuilding their perspectives on time. For, if its reconstitution – consisting in the shift from the mechanical to natural/mythical time perspective – yields effects in terms of blurring the boundaries between the present, the past and the future, then it offers a sort of *tabula rasa* from which one can look at events afresh and either see again or create a new meaning. Thirdly, the future becomes a sort of borrowed time in the sense that it serves as its repository – what is to come is, in the end, a contribution to the present – as well as a safety valve – the sense of "not-the-end" of time leaves room for possible changes.

Fourthly, and perhaps most importantly, what also links these texts is that their authors resolve the same dilemma in a similar manner: the concept of mechanical time that has lost its potential to generate meanings is replaced with natural/mythical time. All writers make their heroes wake up, stand against and/or run away from their hitherto lacking meaning lives and embark on journeys that would bring their lives a sense of meaningfulness. This manoeuvre translates into their views of time. Disappointed with the clock-time perspective, they distance themselves from such perceptions of time and set off on intellectual journeys in search of such concepts of time that would make their lives meaningful. Such journeying translates further into the twentieth century discussion on time and generating meaningful relationships.[53]

Graves's time-and-meaning dependence might be said to consist in a kaleidoscopic understanding of events – whatever happens is understood each time anew with relevance to the surrounding conditions, just like a piece of glass in a kaleidoscope – when it changes its place, it changes its meaning. An event, object or person is obviously always a part of the cycle of growth and death that gives the general frame to the Gravesian kaleidoscope, but it can shift between the cycles of Graves's White Goddess Myth. Should its meaning start to wear out, each shift would provide it with a new meaning. Malcolm Ross sees meaning as moulded on various levels simultaneously. Inasmuch as he locates time and meaning-generating relationship in the same general frame as Graves, i.e. the divinely-orchestrated life of an individual, when it comes to particular objects of interest, they receive their meaning when they appear at points at which temporal layers intersect. Alejo Carpentier completely defies the idea of meaning being moulded by clock time. For him, mechanical time enslaves and makes one's life drained of its significance. Just like Graves and Ross, he opts for nature, where events simply *are* as a part of "the matter of course"[54] imbued with a meaningful potential. Taken together, the time manipulations of these writers might point to a direction that has been taken in the research on the dependency between time and meaning generation: if mechanical laws are insufficient to explain how the social world works, then perhaps a return to the basic and unchanging laws of nature might provide time concepts that would ensure meaningfulness of the real.

References

Bigelow, J., 1996, "Presentism and Properties," in *Noûs* vol. 30, Supplement: Philosophical Perspectives, 10, Metaphysics, Cambridge and Oxford: Blackwell Publishing, 53-52.

Carpentier, A., 1991, *The Lost Steps*, Harriet de Onis, Minerwa, trans., London: Minerva.

Daiches, D., 1965, *The English Novel and the Modern World*, Chicago, London: The University of Chicago Press.

[53] The key works that discuss this relationship are Martin Heidegger's *Being and Time*, as well as Hans-Georg Gadamer's *Truth and Method*.

[54] Carpentier, *The Lost Steps*, 205.

Diana, M., 2012, "Scale and Cognition in Historical Constructions of Space," in *Historein* vol. 10, <http://www.nnet.gr/historein/historeinfiles/histvolumes/hist10/historein10-mishkova.pdf>, 94-105.

Eliade, M., 1959, *Cosmos and History: The Myth of the Eternal Return*, Trask, W. E., trans., New York: Harper and Brothers.

Graves, R., 1975, *Seven Days in New Crete*, London: Quartet Books.

Heidegger, M., "For Edmund Husserl on His Seventieth Birthday, April 8, 1929," ed., and trans., Thomas Sheehan, <http://www.clas.ufl.edu/users/burt/pdfs/e-8-iv.pdf>, 3.

—. 2008, *Being and Time*, New York: Harper & Row.

Ross, M., 1950, *The Man Who Lived Backward*, New York: Farrar, Straus and Company.

Koselleck, R., 2002, *Practice of Conceptual History. Timing History, Spacing Concepts*, Presner, T. S., and et. al, trans., Stanford: Stanford University Press.

CHAPTER EIGHT

"AND NOW TIME HAS MADE ME HIS NUMBERING CLOCK": TIME AND POETRY IN SHAKESPEARE'S *RICHARD II*

JACEK MYDLA

Richard II is one among a considerable number of Shakespeare's plays that can be read as a study of human time, including *As You Like It*, *Macbeth*, *Antony and Cleopatra*, *King Lear*, *Measure for Measure*, *Pericles* and *The Tempest*. These plays address various aspects of our relation to time, the different ways of experiencing the essentially temporal (time-related, time-conditioned) nature of our existence, and the many ways we make sense of and utilise time's three dimensions. These dramatic explorations of human time, as we might call them, also address attempts of the human mind to escape the pressures and constraints of the unstoppable temporality (time's "progress"; the flux) and to find the haven of an eternal present. Moreover and significantly, due to the verbal texture of the dramatic medium, the plays are explorations of the entanglements between human time and language, of how our concerns with time and our temporal negotiations are not simply *expressed in* but also *related to* language, and – because the dramatic use of language foregrounds language's essential interpersonal dimension – to language as a medium of interaction (a mode of existence of the "social animal").

The facets of human time that Shakespeare's dramaturgy explores vary from play to play. To name a few instances: *Antony and Cleopatra* pits two senses of time against one another: time as experienced and consumed by lovers and time as the element that conditions effective operations in the realm of "international" politics; in *Measure for Measure*, pregnancy and breeding are contrasted with infertility and futility, understood both literally and metaphorically; blood relations and continuity make up the joint concerns of *King Lear*; in *Pericles* Shakespeare transcends the tragic sense of cursed or stifled futurity (acutely experienced in *Hamlet*) by opening the plot onto the possibility of rebirths and reunions.[1]

Richard II is a language-focused and a time-focused play; both these foci meet and fuse in scenes and speeches in which it is suggested that time depends on language, that time also can be "performed."[2] Absurd as this idea may at first sound, it does make sense – as I will argue – once we agree that the represented world in drama is a realm of interpersonal relations, and as such verbally malleable.[3] The essential reality portrayed in a play is a reality that "takes place" in and through language; it is negotiable and dynamic. There is no reason why time as a factor in human affairs should not share these features. As one of the characters in *Richard II* puts it with a degree of astonishment, the lengths of time, days and seasons, "lie in words" (1.3.212)[4]; to put this differently, time inheres in language.[5] In this way, Shakespeare teases us intellectually by suggesting this interdependence of time and language; for if humans are essentially conditioned by time, human nature is also related to and conditioned by its capacity for verbal expression and verbal behaviour.

---

[1] For detailed analyses of these plays and these issues see my *The Shakespearean Tide. Studies in the Dynamics of Human Time* (Katowice: Wydawnictwo Uniwersytetu Śląskiego, 2012).

[2] For a speech-acts oriented analysis of *Richard II* see two of my other publications: "Undoing Selves. *Richard II*, Theatrical Incarnation, and Verbal Exile," in *The Writing of Exile,* eds. Wojciech Kalaga and Tadeusz Rachwał (Katowice: Wydawnictwo Naukowe "Śląsk," 2001), 207-223; and the chapter on *Richard II* in *The Shakespearean Tide. Studies in the Dynamics of Human Time.*

[3] What matters in drama is how characters negotiate their mutual relations, and the external circumstances are of secondary importance. The portrayed world is "dramatic" because it is never fixed or stable, but always in a state of flux; it is a product of negotiations and its dynamics corresponds to and reflects the dynamics of interpersonal ("inter-figural") relations.

[4] All my quotations from *Richard II* are from the Cambridge edition by Andrew Gurr (Cambridge: Cambridge University Press, 1992; first published in 1984); I have also consulted the latest Arden edition by Charles Forker (London and New York: Routledge, 2002).

[5] It may be unnecessary to remind ourselves that Shakespeare was neither a linguist, nor a philosopher. However, his work has always possessed the ability to puzzle linguists and perplex philosophers. Not only do many of the plays sustain "scientifically" equipped, scholarly inquiries; they generously repay academic research and have never ceased to inspire and enlighten systematic investigations of problems fundamental to human existence.

Richard II is a poetic drama; it features a protagonist who, to begin with, enjoys the prerogatives of royal speech, and the power that resides and is expressed in a king's language. But, as he is stripped of his power, Richard becomes more and more poetic, until, in his final moments, he is left with language as his solitary companion. This dynamics of Richard's relationship with language is shown in contrast to his political rival, Bullingbrook, who is consistently a man of few words. Richard is in this respect like Shakespeare, who has found figurative language irresistibly attractive. In one of the most poignant dramatic moments in all of Shakespeare, Scene 5.5, we see the confined and alienated Richard whiling away his "empty" time. The scene opens with lines in which Richard attempts some sort of mental acrobatics, the goal of which is to populate "this prison" with thoughts. As we know, he is possibly awaiting execution at the hands of his successor's henchmen. Indeed, murderers enter shortly after he has delivered his soliloquy, which is very long (lines 1 – 66).

The ludicrousness of the first part of Scene 5.5 consists in the fact that *nothing happens*, i.e. not until Richard is killed (the murder "takes up" 7 lines, spoken by Richard: 105 – 112). It seems almost absurd that Shakespeare decided to devote so much stage time to so much time devoid of action. Richard sees more and more clearly that *time* is elsewhere, namely with the new king. The language that Richard is left with, however, is not only a companion with which/whom to fill out solitary hours; Richard uses it as a means of self-searching. Shakespeare makes his fallen hero confront his situation – that of a monarch who has been "unkinged" (4.1.219; 5.5.37) and does not even know what name to call himself (4.1.254 – 258) – by causing him to think about time. At one point, Richard hears a piece of music being performed outside his cell, which makes him reflect upon time, a painful reflection, since now he bitterly feels his powerlessness with regard to time: "time has made [him] his numbering clock" (5.5.50).

What will interest us in this article is how Shakespeare defines – perhaps redefines – the relationship between the poetic use of language and power. How is it at all possible that the unkinged and alienated Richard can and does speak, if – as the first movement of the play shows – effective use of language is made possible by power? To say – because the disempowered and powerless Richard's solitary musings are radically ineffectual – that his speeches *do not perform anything* raises the problem of how is it possible that they still make sense, to him and to the audience? Richard, the unkinged monarch turned poet, certainly uses a language that is a-social and, we might say, "purely poetical"; to use Richard's metaphor from the abdication scene, his speech is no longer "sterling" in England (4.1.263); now he can *do nothing* just by speaking.[6] He seems to be left with a language that is not capable of *performing* anything (a language from which illocutionary power has evaporated). And yet Shakespeare gives his fallen king a language that is both lyrical and ornate. Is then Shakespeare's point to demonstrate that the source of poetry is a crisis of language in the social realm (as in the case of Richard's disempowerment)? We shall see, however, that "authentic" poetry is not merely a means to chase away or "kill" empty time; *Richard II* shows that disempowered language can be a means of inquiring into the nature of human time in its relationship to language, no matter how painful the thus-attained discoveries might be.[7]

No single formula or definition can encapsulate the meaning of human time; consequently, no single play is concerned with only one facet of human time. Not surprisingly, *Richard II* shares a number of themes with other time-concerned plays. There are, however, several features that distinguish *Richard II* even among the plays in which Shakespeare explores human time with special pertinacity. Here we shall look closely at one of them, namely the relationship of time to language. That Shakespeare was intrigued by and drawn to these relations – how time depends on language and how language depends on time – is evident to a reader of his poetry. A recurring idea in the sonnets is that a poem can be an effective means of transcending the limitations that physical or natural time imposes on man and thus of realising our dreams of immortality. In *The Rape of Lucrece*, the title heroine resorts to language with the hope that speaking will help her regain her composure after the sexual assault she has been a victim of, but her lengthy speech *on time* leads her finally to an illumination: speaking has diverted her from what the situation demands of her: suicide as a means of retaining her dignity.[8] Shakespeare's King Richard II, as we have pointed out, enjoys and abuses (in the eyes of his opponents) his sovereignty over time, but his enforced abdication has thrown him into something like a social void; he has

[6] To confirm that Richard's "word" has lost "current value" (Gurr's explanation of "sterling"), it is Bullingbrook who gives the order, not Richard: "Go some of you, and fetch a looking glass" (4.1.267).

[7] In terms of its genre, *Richard II* is described as a tragedy, which makes us expect that the main protagonist, the tragic hero, will attain the state of mind called recognition or illumination (Aristotle's *anagnorisis*).

[8] For a time-focused analysis of *Lucrece* see the relevant chapter in my book *The Dramatic Potential of Time in Shakespeare* (Katowice: Wydawnictwo Uniwersytetu Śląskiego, 2002).

become a nonentity, a condition that he experiences as a temporal stasis.[9] Like Lucrece, forcefully divested of chastity, the "unkinged" Richard has language to keep him company; like Lucrece, he makes time the subject of his solitary and agonising musings.

We have already mentioned the idea of the inherence of time in language. At the end of the "lists" (or, trial by combat) scene (Scene 1.3) – it will be recalled – Richard intervenes and prevents a duel between Bullingbrook and Mowbray and then banishes both contestants. While Mowbray must never return to England, Bullingbrook's exile is to last ten years ("twice five summers," in Richard's phrasing, "summer" standing in a synecdochical relationship to "year"). A moment later, softened at the sight of old Gaunt's (Bullingbrook's father's) grief, Richard shortens Bullingbrook's exile to six years ("six frozen winters"). This causes Bullingbrook to reflect upon the dependence of time on language: "How long a time lies in one little word!" – he exclaims – "Four lagging winters and four wanton springs / End in a word; such is the breath of kings" (1.3.214).

This is a significant moment for our subsequent analysis of scenes later in the play and chiefly of Richard's soliloquy towards the end of the play. Firstly, Bullingbrook's lines, with the help of the synecdoche used earlier by Richard (seasons, with added qualifications suggesting human time, "lagging" and "wanton," stand for the year) make us aware that language can exercise sovereignty over time; royal speech shortens ("ends") time. Secondly, we are simultaneously made alert to what linguists and philosophers of language now call the *performative* use of language. The metonym "breath of kings" points to the king's authority or prerogative, the entire scene being a demonstration of it, a veritable chain of speech acts, the decisive ones performed by Richard. These two observations point to Shakespeare's major concern in the first movement of the play: we see Richard wielding language as if it were a weapon, or, perhaps more appropriately, a sceptre. But we must not forget that scenes like the one we have been discussing prepare us for those in the play's latter movement, in which we see Richard being gradually divested of all manner of authority and sovereignty, indeed of his identity.

Before we turn to the scenes that depict Richard's downfall with the aim to look closely at the (different?) functioning of language in them, we need to emphasise one more aspect, namely the already-mentioned figurative use of language, chiefly by Richard, but by no means

[9] Richard obsessively returns to the idea of nothingness: "for I must nothing be" (4.1.200); also in his last soliloquy: 5.5.36 – 41.

exclusively by him. It is, of course, a general characteristic of the way in which Shakespeare addresses issues – and there are many of them – which we may describe as existential (human time among them) that he makes his characters use figures of speech much more liberally than a philosopher would. The lines just analysed show that Shakespeare not only goes beyond concrete time (the so-called temporal deixis),[10] but also "embellishes" political and "legalistic" usage (Richard's decrees and the responses they elicit). On the one hand, Shakespeare never allows us to forget that he is a poet, and, consequently, all his characters – no matter how petty, pedestrian or perverse – are rich inheritors of and live testimonies to his talent.

In the context of *Richard II*, this commonplace of Shakespeare scholarship, acquires special significance, and this for two reasons. Firstly, Shakespeare has devised and written the play in such a manner that it heightens our awareness that we are speaking animals. Speech defines us as human beings; the metonym "breath" for "speech" (occurring again and again as used by *various* characters; e.g. 1.1.173, 1.3.153, 1.3.172-173, 2.1.1 and 8, 3.1.20, 3.2.56 and 164, 3.3.33, 3.4.82, 4.1.209) is clearly a means to this end. As we have also observed and will inspect promptly, *Richard II* has a language-conscious protagonist, a king who, stripped of sovereignty, has become a poet. Language keeps Richard company literally until he "breathes" his last words.

In the context of drama, however, figurative or "poetic" language (so-called imagery) is not to be understood in the "romantic" meaning of poetry as a pouring forth of feelings (Wordsworth's "spontaneous overflow of powerful feelings"[11]). Rather, figures of speech such as those we have just looked at in scene 1.3 of *Richard II*, fulfil functions that may be called rhetorical.[12] Rhetoric in its practical aspect is any speech in pursuit of interests, and this understanding of rhetoric can be linked with

---

[10] References to the calendar, to hours and days, are used chiefly for mimetic purposes; temporal deixis has the aim to produce a sense of reality (to enhance verisimilitude). Besides this function, concrete short-term references to the future ("tonight," "tomorrow") may also raise and sustain suspense. See the relevant sections in my *Dramatic Potential of Time*.

[11] The famous passage from the Preface to *Lyrical Ballads* reads: "I have said that Poetry is the spontaneous overflow of powerful feelings: it takes its origin from emotion recollected in tranquillity … ." William Wordsworth, "Preface to *Lyrical Ballads*," in *The Norton Anthology of Theory and Criticism*, ed. Vincent B. Leitch (New York and London: W. W. Norton & Company, 2001), 661.

[12] I analyse this at more length in the "Introduction" in *The Shakespearean Tide. Studies in the Dynamics of Human Time*.

the idea of performative language as worked out by speech act theory. Characters in a play primarily use language, not to express how they feel, but to act.[13] In other words, "poetry" in drama is not to be understood as lyric but as verbal doing ("doing things with words," in John Austin's famous phrase[14]).

This understanding of "rhetoric" in the sense of verbal doing does not violate the time-honoured and, indeed, common designation of the term: art of persuasion. More generally, rhetoric in the context of drama consists in the use of figurative language, not (at least not primarily)[15] for the purpose of constructing a fictive world or expressing the emotional states of the characters, but in order to shape and reshape relations between the characters and in this way to direct the course events. Poetic imagery, when it occurs in drama, unavoidably assumes a dynamics: its purpose is chiefly to shape an interpersonal situation and to give it a direction. Imagery that is used to "describe" time is no exception; time-related images are also rhetoric in the sense just stipulated. This is to say that the rhetoric of time in drama is dynamic, and future-oriented.[16]

It should be obvious by now that figurative language is not impersonal; in drama, characters use imagery to express attitudes and to exert influence over other characters. This permeation of language by interest (the way in which verbal interaction can be seen as a power struggle) is a characteristic of

---

[13] "Acting" in the sense of spoken action, as applied to drama (a dramatic situation), has two basic senses: 1) self-determination (a decision to do this or that: Othello wants to kill Desdemona "this night") and 2) influence (an attempt to make *another* character do this or that: Desdemona pleads with Othello to let her live). These two types of verbal behaviour make up what can be called the *dramatic* function of language; without them, there is no drama.

[14] This reference is to the title of John L. Austin's book *How to Do Things with Words* (Oxford: Clarendon Press, 1962); see *The Norton Anthology of Theory and Criticism*, ed. Vincent B. Leitch, 1430.

[15] Poetic language, for instance metaphors, can of course be used "mimetically," that is, to build up and enrich representations of time as an entity at work amidst human affairs. The phrases that immediately come to mind are obscure and perplexing: e.g. "the seeds of time" in *Macbeth* and the disjointed time in *Hamlet*.

[16] The question that arises at this point is this: since every image is equipped with a dynamics, can there be imagery that is not directly or indirectly related to time? The answer must be negative. There are of course degrees of indirectness. Often in Shakespeare the imagery used to speak of time is "natural," "organic" or "biological"; this is also how we conceptualise (and visualise) time in real life. In *Richard II*, this imagery is used by the Queen, who thus expresses her anxiety: "Yet again methinks / Some unborn sorrow ripe in Fortune's womb / Is coming towards me …" (2.2.9 – 11).

the non-literary functioning of language, and in this very general sense, drama, regardless of the degree of poeticality, imitates or, to use a better word, reflects the verbal behaviour of real people. Little wonder that in Shakespeare it is paramount. To simplify matters we could say that language is a mere instrument; but this would be an unfair simplification. If people are (defined by) what they feel and what they wish to achieve, then to the extent to which language is capable of expressing these two sides of human nature (of being human), *people are language*. Language – and thus also language as it is used in drama – is neither objective nor subjective; it is *personal*, and thus also and by implication *inter-personal*.[17]

It will have become clear by now that the elements of the theory of drama as presented so far are rooted in the basic insights of the speech-act theory of language (or, more, generally, pragmatics). Performative pragmatics posits that "reality" is co-produced by language, and is, to this extent verbal. In this sense and to this extent, it is needless to insist on a separation between reality and drama. The reality in which we *as people* participate is the same in nature as the fictive world represented in a play: this reality and that world are social constructs; for their fabric, they have personal interactions and their dynamics depends on the pursuit of goals and interests. In both reality and drama, language is the essential mediating substance.[18]

The so-called solar imagery in *Richard II* is a relevant example of the dynamics of imagery as it is used in drama. It is eagerly embraced by Richard, who – despite fresh news of Bullingbrook's rebellion and of his growing strength "in substance and in power" (3.2.35) – finds in it assurance of his immunity: "So when this thief, this traitor, Bullingbrook … / Shall see us rising in our throne the east / His treason will sit blushing in his face / Not able to endure the sight of day, / But self-affrighted tremble at his sin" (3.2.47 - 53). Soon, however, Richard will have to face

[17] Often *time-related* imagery is used to express attitudes and to reveal expectations and purposes. In *Much Ado About Nothing* Claudio uses a personification of time to express the impatience of a young man awaiting the wedding ceremony: "Time goes on crutches till love have all its rites" (2.1.334). In *King Lear*, Cordelia also suggests a personification, linking it with a revelatory function: "Time shall unfold what plighted [i.e. hidden] cunning hides" (1.1.274). In both these instances, typically, the imagery has a future-oriented dynamics.

[18] In this analysis, we have narrowed down the idea of communication to its verbal component. This is because of the nature of the literary material in hand; Elizabethan drama is spoken drama. It is obvious that non-verbal communication must not be ignored in a comprehensive theory of drama.

the unwelcome truth and change his rhetoric accordingly. His is not Sun, after all; rather, like Phaeton known from Ovid's *Metamorphoses*, he has been a reckless and heedless charioteer. Now time has come for him to step down and see the rise of his rival (3.3.178). In the abdication scene, the crying Richard calls himself "a mockery king of snow," who melts "before the sun of Bullingbrook" (4.1.260). This disowning of the solar imagery completes the process of stripping Richard, not simply of power but of identity.

As the intertextual link with the legend of Phaeton suggests, the solar imagery is not only potentially dynamic but is also charged with moral connotations. For Richard – initially believing that language, and by consequence time as well, belongs to him – this imagery has turned out to be a trap. As a consequence of his (royal) appropriation of the performative properties of language, he has identified with time; his eager adoption of the Sun imagery makes this clear, as does Bullingbrook's line about the powerful "breath of kings." But the Sun-king association is misleading because it is based on the idea – indeed, a misapprehension – that kingship (the office of being a king as part of the country's constitution) – like the Solar System – lies outside language, outside the area or arena of human interaction (with its inalienable flexibility and dynamics), and thus cannot be affected by human agency.[19] The lesson that Richard has learnt is that nothing in the interpersonal realm is non-negotiable; to the extent that human time depends on language, time is also a negotiable entity.

The deposition (or abdication) scene, Scene 4.1, in which Richard is "unkinged," is proof that he has been wrong all along, that he has failed to grasp the true nature of kingship, the way it is essentially related to how language functions in the arena of human interactions (to the performativity of a person's social identity). He has been wrong in not taking into account the negotiability of kingship. Shakespeare's handling of the historical material suggests that in this misconception, in this belief in the absolute authority of "an anointed king" (3.2.55) lies the nature and cause of

[19] For this reason, any sharp distinction between different types of constitutions, as e.g. between the monarchical and republican systems, does not make much sense. What political philosophers have called "the social contract" is indeed at the bottom of any system; and contracts are negotiable. This explains why Queen Elizabeth found the scene of Richard's abdication treasonous and had the performances suppressed. This of course also explains why Elizabeth's opposition used that same scene to convey their message. (For the performance history of the play, see the introductions to the Arden and Cambridge editions.)

Richard's downfall, rather than in misgovernment.[20] Appropriately, at the end of the abdication scene, when Bullingbrook gives orders to "convey" Richard to the tower, Richard accuses all those present of conspiracy against him: "Conveyers are you all / That rise thus nimbly by a true king's fall" (4.1.315 – 316). Without a conspiracy of many, Bullingbrook would not have become king. It is this compact, this "conspiracy," that has stripped Richard of his power and name; it is this compact that has left Richard's time void of all substance and meaning.[21]

In Richard's final scene, we hear his soliloquy (or rather two soliloquies, for the first one is interrupted by music) and then see him butchered. Represented as a poet, alone and alienated, Richard is speaking to himself.[22] Admittedly, he is doing so in order to alleviate the pains of his predicament and to chase away the tedium of solitary confinement. His language should be entirely expressive, i.e. poetic or lyric. Because of the imprisonment, almost all interaction has become impossible. We should expect – and do witness here – a verbal crisis, an implosion of meaning. Richard knows – indeed seems have an acute sense of this – that what he is saying cannot mean anything to anyone. Now he has no "audience" (in the sense of a host of obedient subjects) over whom to exercise his authority, something that *the former King Richard* enjoyed immensely. Now – despite the moral disapprobation that may have built up in us (e.g. due to his verbal abuse of the dying Gaunt in 2.1) – we are free (perhaps the playwright expects us) to sympathise with the abandoned and humiliated man, who still remembers what it felt like to be a king.[23]

---

[20] The play is not consistent in this respect. Richard's opponents represent him as an improvident and imprudent monarch. Richard accuses himself of having "wasted time" (in the final soliloquy; see below). The whole play does not entirely confirm this portrait; the plot singles out the banishment of Bolingbrook and the consequent rebellion as the chief causes of Richard's downfall.

[21] Richard has become ~~Richard~~. See my "Undoing Selves. *Richard II*, Theatrical Incarnation, and Verbal Exile." See also the self-annulling lines 3.2.175-177, the conclusion of one of the most moving speeches in the play (lyrical, due to the amount of self-pity).

[22] The circumstances bring to mind Percy B. Shelley's metaphor: "A Poet is a nightingale, who sits in darkness and sings to cheer its own solitude with sweet sounds ... ." See his "A Defence of Poetry," in *The Norton Anthology of Theory and Criticism*, ed. Vincent B. Leitch, 699. But there are differences: Shelley's poet has "entranced auditors" and, besides, Richard's reflections are not sweet but bitter.

[23] The play is ambiguous also with regard to what response in the audience it elicits. One the one hand, Richard overindulges in self-pity, even though the speeches that punctuate the stages of his downfall are wonderfully poetic and

Richard's musings are interrupted by music. Stage directions (e.g. in the Cambridge edition[24]) read: "The music plays." This is a means to make Richard start thinking about time: "Music do I hear? / Ha, ha, keep time! How sour sweet music is / When time is broke and no proportion kept. So is it in the music of men's lives." (5.5.42 – 44). In this simple and yet apt and skilful way, Shakespeare redirects the course of Richard's thoughts and introduces the ideas of rhythm and harmony ("proportion," order), and also, consequently, of time. Musical performance is about "keeping time" and people have the natural capacity (Richard's "daintiness of ear") to appreciate regularity and to detect disorder and disharmony; "to check time broke in a disordered string" can be paraphrased as meaning "to detect lack of rhythm and harmony."[25] The phrase "the music of men's lives" metaphorically transposes the impersonal meaning of time (an "ideal" sense and flow of time as embodied in music, if properly performed) to the sphere of human interactions and the moral meanings attached to them. How is it possible, he asks, that he can "keep time" in his mind (which makes him capable of detecting a broken rhythm), despite having, metaphorically speaking, lost the sense of rhythm or harmony in his life and become a "broken" man?

We could, perhaps, say that the device Shakespeare has decided to use here is similar to torture. The piece of music that Richard overhears certainly works like an irritant. At the end, Richard exclaims in anger that it "mads" him: "This music mads me" (5.5.61). For one thing, its lack of orderliness (failure of the musicians to keep time) causes him to reflect upon his disorderly life: he has not been able to hear "[his] true time broke." The reason for this anger is not the poor performance. The music has made Richard realise that he knows what it means to "keep time" in life and yet has not been able to manage his life and his office. The poorly performed piece of music irritates him because he projects the story of his downfall onto it.

The second part of Richard's soliloquy in the final scene has typically for its main motif a personification of time. Somewhat absurdly, this personification *de*personalises Richard. Not only does Richard find himself in time's power; he has become something like a living clock: "I

---

memorable. On the other hand, we cannot take the side of his opponents as wholeheartedly as in the case, say, of *Richard III* and *Macbeth*.

[24] Stage directions in the contemporary editions of Shakespeare depend on the "original" editions consulted; it will be remembered, however, that they are of secondary importance because the text of the play itself is the main source of stage directions.

[25] See relevant editorial matter in Forker's edition (465).

wasted time and now doth time waste me, / For now hath time made me his numbering clock."[26] In this way, Richard describes himself as a time-piece ("numbering clock"), an instrument that measures time's progress but which in itself is devoid of any substance. The insight is basic: *human time cannot be measured* ("numbered"). Richard's sense of having become something like a mechanism is a consequence of his alienation. Given the political context, this makes perfect sense: the centre of authority has shifted, away from Richard and to Bullingbrook. "His" time – somewhat ridiculously but to the point – is shown as Bullingbrook's servant. "My time," says Richard, "runs posting on in Bullingbrook's proud joy …" (5.5.59).

Richard has an acute sense of the slowness of his time, of the tedium of being an instrument that is doomed to measuring the flow or current without being able to enter the stream. "Running" time is live time, as opposed to stagnation and inaction. Real or filled time is now in the service of the new king, eager to keep up with events as they are shaping themselves in the dynamically changing interpersonal realm (with King Henry in its centre). The imprisoned Richard, Bullingbrook's "Jack of the clock,"[27] is trapped in a time-sense that feels like timelessness (paradoxically, because of the orderly, even flow of the "numbered" time!). Language may be used by the prisoner to entertain himself, but if this happens, the "endless now" will be extended. Language (and the "entertainment" it is capable of providing; how much time we can "kill" by speaking) may help Richard to attain temporary oblivion, but he will never be able to fill his time with substance, i.e. to recreate a world that would be a substitute for his lost kingdom, populated as it was by obedient subjects. The bauble of oblivion which *Richard* has been at pains to produce and inflate with the aid of poetry, *Shakespeare* has now punctured, as it were, by music, a moment suggesting that an awareness of the passing of time – the temporal apperception as constitutive of the human condition – cannot be stifled for long.

But what exactly does Shakespeare mean by *broken* human time as experienced by Richard? Are we to conclude that the "poetic phase" of Richard's downfall is equal to an admission of failure? Is his fifth-act

[26] Much earlier Richard reflects: "… time has set a blot upon my pride" (3.2.81).

[27] Gurr in the Cambridge edition explains: "figure which strikes the bell to sound the hours" (168); Forker supplies an elaborate elucidation: "the small mechanical figure on certain old clocks of a man who struck the bell with his mallet to mark the hours or quarter-hours. Richard complains that for the politically triumphant Henry [Bullingbrook] time passes rapidly and joyfully, whereas he himself must eke out his poor existence hour by hour in prison" (467).

soliloquy nothing more than an expression, a *verbal symptom* of his loss of all power and of socio-political erasure? Is the goal of his elaborate if painful rhetorical performance to speak himself into non-being? Has he become a voice without a speaker? Once a powerful weapon, language has now become a sword with a blunted blade, a tool with which to commit a masochistically cruel suicide. Yet, are we really witnesses to how Richard is being "wasted" by time?

The first part of the soliloquy is very different from the second. While in the first part Richard makes an attempt to create a fictive kingdom and people it with his thoughts (as a pitiful replacement for actual subjects), in the second he diagnoses and dissects his condition. In the first part, he attempts to "kill" time, to entertain himself; in the second, the awakened and live experience of time (brought about by the overheard music) makes him confront the essentially temporal dimension of human (and of particularly his) existence. In the first part, he uses language instrumentally, i.e. as a tool with which to chase away tedious moments; in the second, language is the dissecting knife. This explains why Shakespeare has decided to offer his hero a long moment of self-searching. Time-killing would be pathetic, even ridiculous; but the actual sober examination (assisted and made possible by time/music and figurative language) elicits our sympathy, perhaps even tinted by admiration.

In the second part of his soliloquy, Richard uses the idea of time to shape his plight into some meaningful whole. We see him use time (his "empty" time) to create a poetic performance that has a substance and a meaning. No matter how distressing the conclusions he reaches may be, he has been able to clothe his predicament of "wastedness" with a cloak of intelligibility. After all, then, Richard is not merely a poet; he is still a doer, a performer, albeit now a humbled one, a speaker who has learnt that language is not a sceptre obedient to his command and that time is not one of his subjects. And yet, his lone-speaking (his soli-loquy) and the illumination it produces, that of the Aristotelian tragic recognition, is – was – for Shakespeare a way to offer his protagonist, no matter how downtrodden and humiliated, a *moment of clear vision*.

Death is approaching Richard in a double sense: his murderers are on their way; the playwright is far in his fifth act and *Richard II* is a tragedy. He is doomed. But "Death" (yet another image of Time) can be merciful; to use Richard's own words, he has been allowed "a breath, a little scene, to monarchize." Breath (language) and scene (time) have conspired to redeem Richard's sovereignty. Shakespeare has indeed created for the unkinged Richard a space and a moment in which "to monarchize," an opportunity which may have been denied the actual king.

References

Leitch, V. B., et al., eds., 2001, *The Norton Anthology of Theory and Criticism*, New York and London: W. W. Norton & Company.

Mydla, J., 2001, "Undoing Selves. *Richard II*, Theatrical Incarnation, and Verbal Exile," in Kalaga, W., and Rachwał, T., eds., *The Writing of Exile*, Katowice: Wydawnictwo Naukowe "Śląsk," 207-223.

—. 2002, *The Dramatic Potential of Time in Shakespeare*, Katowice: Wydawnictwo Uniwersytetu Śląskiego.

—. 2012, *The Shakespearean Tide. Studies in the Dynamics of Human Time*, Katowice: Wydawnictwo Uniwersytetu Śląskiego.

Shakespeare, W., 1992, *Richard II*, Gurr, A., ed., Cambridge: Cambridge University Press.

—. 2002, *Richard II*, Forker, Ch., ed., London and New York: Routledge.

CHAPTER NINE

"DREAM FILMS": WOMEN'S TIME AND SLEEP

KAREN HEALD

The Chora

The *Chora* sets the context for a "psychic space" relating to the condition and existential situation that I am researching from an artistic perspective. It begins from my encounter with Kristeva's terms "semiotic *chora*"[1] and "women's time"[2] alongside my original concept of "transitory strata."[3] In relating the terms semiotic and *chora*, Kristeva argues that the

---

[1] Julia Kristeva's notion of the "semiotic *chora*" and time, in relation to linguistics and feminist critical theory, acts as a model for this research. In the first place, the terms semiotic and *chora* frequently remind us that the *chora* is the space in which the meaning that is produced is semiotic. Kristeva uses maternal language when discussing the *chora*, speaking of its motility and comparing it to contractions and the maternal part of giving birth and the birth itself. She describes the "'energy' charges as well as 'psychical' marks, [which] articulate what we call a *chora*: a non-expressive totality formed by the drives and their stases in a motility that is as full of movement as it is regulated and cannot be represented." Julia Kristeva, *Revolution in Poetic Language* (New York: Columbia University Press, 1974/1984), 25.

[2] In "Women's Time", Kristeva discusses women's link with cyclical and monumental time as a conceptualising of time from the context of reproduction and motherhood, and the need for a future generation of feminists to confront the task of reconciling maternal time (motherhood), with linear (political and historical) time - readily labelled masculine. Julia Kristeva, "Women's Time" in *New Maladies of the Soul* (New York: Columbia University Press, 1979), 201-224.

[3] "''Transitory strata'' are preverbal, personal spaces which draw from Kristeva's semiotic "chora" and, through the making of experimental "dream films", visualize a re-interpretation of sleep, the unconscious and experiential time. This differs from the split or the divided self as it pertains to fleeting parallels of time that exist

chora is the space in which the meaning that is produced is semiotic, such as: the sounds of an infant's cooing; the psychotic's inability to communicate language in a meaningful way; or the artist who mobilizes the semiotic for creative and novel means. These meanings are significant and are expanded on in "transitory strata," a term which describes preverbal, personal spaces, that draw from Kristeva's "semiotic *chora*" and through the making of experimental "dream films," visualize a re-interpretation of sleep, the unconscious and experiential time. The "dream films" are key elements and through analogies and cross-readings between theory and practice, I have artistically and collaboratively worked through this research. The *chora* has enabled me to explore and expand on preverbal spaces by the use of subtle suggestions within the films. I have used this approach as a method for creating artworks and using video as a creative tool. The discourses and implications around the *chora* have assisted me in articulating what I am seeking in the artworks/practice. In addition to "transitory strata" I discuss two further key terms of mine – "being in-between" and "in-between-ness." The first is a space that oscillates between the personal preverbal "transitory strata" and the public places of "in-between-ness." The latter is a conceptual, reflective public space offering a synthesis of collaborative arts and science practice.

Although Kristeva has written on numerous academic subjects, in this research I explore Kristeva's texts with an emphasis on cyclical, maternal and monumental time. For this, I draw on my notion of time as being a time and space of "being in-between" different levels of consciousness and of being in simultaneous worlds. I attempt to create the paradoxical – the "unrepresentable" or the "unspeakable" which is addressed in the "unconscious thetic" while opening up for a re-presentation of temporal experiences. The "unconscious thetic" refers to the raw primitiveness of the mind's sensations to create a fluid sense of the self and environment offering an innate and intrinsic act. This attempt to offer another translation of time is achieved through "dream films" where time is proposed as a mechanism in

within the creatives' or the depressives' state of being. This ephemerality is discovered within the individual, lying amidst the layers of consciousness, the realms of waking, reverie and wakefulness. These experiences seldom occur whilst asleep, but can be stirred by direct sensitivity to sounds, which when heightened under extreme moments of creativity, stress or illness, can take a person on a more profound pathway. These journeys, or "transitory strata", have the ability to build and fragment, and the "dream films" aim to echo, the unarticulated and the unutterable." Karen Heald, ""'Dream Films" and Research as Collaborative Practice in Contemporary Art and Science Methodologies" (PhD thesis, Leeds: Leeds Metropolitan University, 2014).

which to devise significant artistic methodologies on the "in-between-ness" of film and research as collaborative practice. By using the camera and editing in this unique way my aspiration for the viewer is that they will experience the "dream films" through multiple layers. It is a paradoxical undertaking to give voice to the unutterable, or perhaps rather to allow it to disperse more widely with a sense of authority and agency. Therefore, my intention for the "unconscious thetic" in relation to editing and the viewer's situation is extremely important.

Kristeva describes two temporal splits in her theory of time – the time of linear history and monumental time – where she links cyclical and monumental time with female subjectivity. She asks for a need for future generations of feminists to confront the task of reconciling maternal (motherhood) and linear (political and historical) time.[4] I am especially interested in this motility and the spontaneous movement between the semiotic and the symbolic, recognising that the *chora* is not just a space, such as the womb, or receptacle but one that precedes language. Plato, from whom Kristeva borrowed the term *chora* relates it to a cosmic motion.[5] I, however, became increasingly drawn to Kristeva's concept of the *chora* and the "psychic" environment of the infant towards the mother's body. Distinct from Kristeva's work which focuses on people with borderline personality disorders, my explorations synthesize the *chora* and its ambiguities between time and film within contemporary art practice. I am interested in this "third space" and it is in the "transitory strata," through the language of fine art, specifically "dream films," where the preverbal personal space of the creatives' or depressives' state of being can encounter momentary parallels of time that are less "beyond us" than "in-between."

As such, non-linear narratives are prevalent in my "dream films," where I explore temporal modes and engage with other female directors working with cinematic time. I explore their relationships with the sense of the passage of time. I seek the experiential quality of the duration of time and the didactic between past and future. Jane Campion's *The Piano*

---

[4] Julia Kristeva, "Women's Time," in *New Maladies of the Soul* (New York: Columbia University Press, 1979/1984), 201-224.

[5] Kristeva extracts the term semiotic *chora* from the Greek philosopher Plato's text *Chora* See Plato, *Timaeus and Critias*, trans. Desmond Lee and Thomas Kjeller Johansen (London: Penguin Classics, 2008). Plato expresses his theories on how the universe was created, describing the *chora* as a space, meaning both a receptacle and nurse, that is, the container and the producer, of what the universe is before anything else exists.

(1993)[6] Stephen Daldry's *The Hours* (2003)[7] and various literary and film critiques are studied in relation to Kristeva's *chora*. I was intrigued by how the tensions between the semiotic and symbolic language in *The Piano* were heightened through music, a path that Kristeva took by introducing music therapy in her psychoanalytical practice. Trinh T. Minh-ha, in the film *The Fourth Dimension* (2001)[8] explored 'tenseless' time in a wonderfully poetic portrayal of contemporary Japan, depicting the image as 'time light.' Minh-ha connects between self and other, human and machine, viewer and image, real and fiction, exploring the connection where past and present could occur. Her films venture into a world of "in-between" realities; into and out of the space of "awakened dreams." This rhapsody of "beyond words and beyond time" could also encompass *The Hours*, relating it to Virginia Woolf's novel *Mrs Dalloway* (1926).[9]

Whereas Kristeva focused on the materiality of language, its sounds and rhythms rather than its communicative element, I focused on the materiality of film via colours, audio/visual properties and its visceral qualities. Sound, colour, the triptych and editing have, therefore, all added to the artistic emanation of the *chora* effect. For example, *pra II,* (2005)[10] explored the relationship between video and painting pictorially within a framed composition and evolved into a dream-like quality through a "small" drama, with the tension heightened from sound by collaborator Susan Matthews. As such, it differed from the mainstream formats of commercial and documentary films (Fig. 1-1).

Another space and time compounded by colour was explored in the film *Paper Interior* (2013)[11] which had a connection with Johannes Vermeer's use of interiors and lighting (Fig. 1-2), whilst *The Artists' Creation* (2009)[12] was a "painterly reverie" that communicated difficult social issues with subtle visual stanzas (Fig. 1-3).

---

[6] *The Piano*, directed by Jane Campion (Australia: Cibey 2000 Home Entertainment, 1997), DVD.

[7] *The Hours,* directed by Stephen Daldry (United States: Paramount Pictures and Miramax Films Home Entertainment, 2003), DVD.

[8] *The Fourth Dimension*, directed by Trinh T. Minh-ha & Jean-Paul Bourdier (Japan: Women Make Movies Home Entertainment, 2001), DVD.

[9] Virginia Woolf, *Mrs Dalloway* (London: Penguin Classics, 1926).

[10] *pra II*, directed by Karen Heald (Czech Republic/UK: Karen Heald , 2005), Video: SD.

[11] *Paper Interior*, directed by Karen Heald & Susan Liggett (Wales: Karen Heald, 2013), Video: HD.

[12] *The Artists' Creation,* directed by Karen Heald & Susan Liggett (Wales: Karen Heald, 2009), Video: HD.

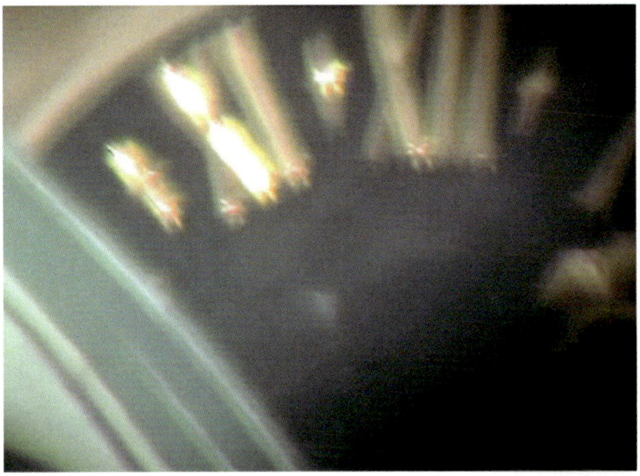

Fig. 1-1: Heald, K. (2005) *pra II* [film still], Testbed 3, Leeds Metropolitan University, Gallery, Leeds, UK.

Fig. 1-2: Heald, K. & Liggett, S. (2013) *Paper Interior* [filmed performance/installation], Wrexham, Wales.

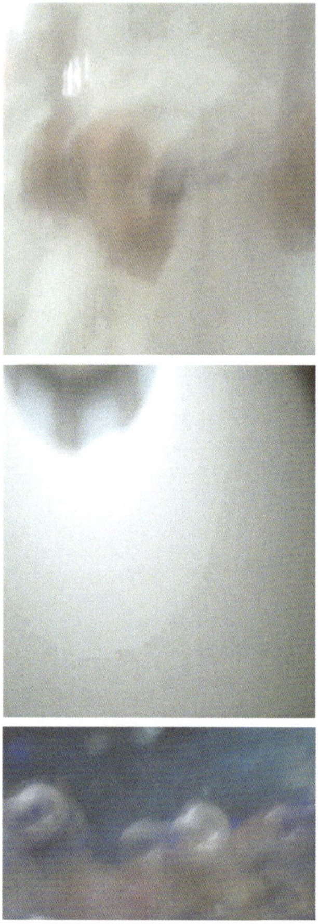

Fig. 1-3: Heald, K. & Liggett, S. (2009) *The Artists' Creation*, [filmed performance], Wrexham, Wales.

I examined the triptych – the fundamental language for the illiterate within religious iconography – and created two different film triptychs. The first one was *Drifting on a Pink Cloud* (2007)[13] which made the audience question where to look. However, this initial work did not hold a rational logic with only time as duration acting as a unifying source (Fig. 1-4).

[13] *Drifting on a Pink Cloud,* directed by Karen Heald & Rachael Kearney (Japan: Karen Heald, 2007), Video: SD.

Fig. 1-4: Heald, K. & Kearney, R. (2007) *Drifting on a Pink Cloud* [photographs from video performance], Kanazawa, Japan.

In contrast, the second film, *White* (2013)[14] had an aesthetic balance of contrast and harmony and explored time and the politics of location. In *White* the triptych acts as a visual discourse and relates three main temporal linear distinctions of the past, the present and the future. The film is layered rather than positioned side by side; there is an inner core of listening, ruminating from the haunting lyrics by Hannah Atkins. The film appears as a mise-en-scene with the pace slowed down and repetition creating a form of reverie (Fig. 1-5).

[14] *White*, directed by Karen Heald & Susan Liggett (Wales: Karen Heald, 2013), Video: HD.

Fig. 1-5: Heald, K. & Liggett, S. (2009) *White* [film still], Wrexham, Wales.

The collaborative work with the dancer and choreographer, Caroline Sabin is cited as an example of the tactile and kinaesthetic relationship to the Kristevan *chora* through the use of the camera and editing techniques. In *Not still at home: poise* (Heald & Caroline Sabin, 2010)[15] Sabin, who is used to choreographing for hours, days or even months before shooting, was filmed spontaneously and impromptu using a fixed focus lens, long take, for 10 minutes. The footage was framed low to the ground and restrictive, which created a performance task for Sabin as a mover. Here I attempted to escape the conventional visual language (Fig. 1-6).

[15] *Not still at home: poise*, directed by Karen Heald & Caroline Sabin (Wales: Karen Heald, 2010), Video: HD.

Fig.1-6: Heald, K. & Sabin, C. (2010) *Not still at home: poise* [video performance], Cardiff, Wales.

Similar to Maya Deren's film *A Study in Choreography for the Camera* (1945)[16] where she uses symbolism in unconventional ways by the method of cutting up images; I rethink how we see things through the camera lens. Deren's film celebrates the principles of the power of movement through extensive editing. In *Not still at home: equilibrium* (2010)[17] the cameras act as the editors through their orchestrated placement, and open LCD screens, creating a montage which was synchronized in the mirror (Fig.1-7). The relationship between both films is that they could only exist as films with the dancers moving in time and space. Along the same lines, Adrienne Rich (1984) discusses the need not to transcend but to reclaim women's bodies.[18] I relate this to Sabin's tactile space and the relationship between her home and herself, speaking of the slippage between transcendence and reclamation.

[16] *A Study in Choreography for the Camera*, directed by Maya Deren (United States: Howard Gotlieb Archival Research Centre, 1945), Film: 16mm.

[17] *Not still at home: equilibrium,* directed by Karen Heald & Caroline Sabin (Wales: Karen Heald, 2010), Video: HD.

[18] Adrienne Rich, *Notes Towards a Politics of Location in Blood, Bread and Poetry: Selected Prose 1979-85* (New York: W. W. Norton & Company, 1984), 210-231.

Fig. 1-7: Heald, K. & Sabin, C. (2010) *Not still at home: equilibrium* [video performance], Cardiff, Wales.

Historically, feminist avant-garde filmmakers have most consistently engaged with time and the unconscious by incorporating psychoanalytic readings of women and their dreams. Early choreographed examples are Deren's *Meshes of the Afternoon* (1943)[19] and *Ritual in Transfigured Time* (1946)[20] which subverted Freudian symbols and sourced Greek mythology to readdress female subjectivity and the human psyche. Through the creation of new images and complex editing techniques Alison Butler (2007)[21] notes that Deren's films poetically point to scientific and technological changes having consequences throughout culture. This is

[19] *Meshes of the Afternoon,* directed by Maya Deren & Alexander Hamid (United States: Howard Gotlieb Archival Research Centre, 1943), Film: 16mm.

[20] *Ritual in Transfigured Time*, directed by Maya Deren (United States: Howard Gotlieb Archival Research Centre, 1946), Film:16mm.

[21] Alison Butler "Motor-driven metaphysics: movement, time and action in the films of Maya Deren," *Screen* 48 (2007), 1-23.

evident in the more recent psychological film *Black Swan* (2010)[22] which depicts a study of a female breakdown unfolding on the screen. The significance of the maternal and the pre-Oedipal in the psychological make-up of subjectivity has been an area that Kristeva has repeatedly returned to, being interested in the very early infancy stage. Here she embraces Melanie Klein's notion of the maternal[23] which precedes Sigmund Freud's Oedipal complex[24] and Jacques Lacan's mirror stage.[25] Kristeva emphasizes that the maternal function and its importance in the evolution of subjectivity gives access to culture and language. Researching psychoanalytical theories and Kristeva's concepts, Laura Mulvey wrote her influential essay "Visual Pleasure and Narrative Cinema" (1975) which evolved feminist theory.[26]

Kristeva discusses Charlotte Perkins Gilman's novel *The Yellow Wallpaper* (1892), which was a response to her neurasthesia – equated today as postnatal depression – as a new feminism.[27] It inspired Patricia

---

[22] *Black Swan*, directed by Darren Aronofsky (United States: Fox Searchlight Pictures Home Entertainment, 2011), DVD.

[23] Melanie Klein studied Freud's writings and developed her own work on childhood theory. Richard. L. Gregory *The Oxford Companion To The Mind* (USA: Oxford University Press, 1987).

[24] Sigmund Freud based his theory on the myth of *Oedipal Rex*, as told by Sophocles in his Theban Plays where Oedipus awakens to his patricide and the unconscious incest with his mother. Anne Baring & Jules Cashford *The Myth of the Goddess* (London: Penguin Books, 1993), 309.

[25] Jacques Lacan, the French psychoanalyst, defines the mirror stage as the moment when a child, seeing [sic] one in a mirror, is enabled to a unitary body image. The screen (ecran) between the subject and the real is part of a specular structure that constitutes the subject. Maggie Humm, *The Dictionary of Feminist Theory* (Prentice Hall/ Harvester Wheatsheaf: Hemel Hempstead, 1995), 173.

[26] Laura Mulvey's article was influenced by the theories of Freud and Lacan, and was one of the first major essays that helped shift the orientation of film theory towards a psychoanalytic framework. Laura Mulvey "Visual Pleasure and Narrative Cinema," *Screen* 16 (1975), 6-18. Prior to Mulvey, film theorists such as Jean-Louis Baudry and Christian Metz incorporated psychoanalytical concepts in their theoretical accounts of the cinema. Mulvey's contribution, however, inaugurated the intersection of film theory, psychoanalysis and feminism. Christian Metz, *The Imaginary Signifier: Psychoanalysis and the Cinema* (Indiana: Indiana University Press, 1982).

[27] *The Yellow Wallpaper* was a 6,000-word short story, first published in January, 1892 in *The New England Magazine*, America, 647-657. It is regarded as an important early work of American feminist literature, illustrating attitudes in the nineteenth century toward women's physical and mental health. Kristeva defines Charlotte Gilman Perkin's text as the new feminism, describing it as a language

MacKinnon-Day's work *Enclosure for Naked Spaces* (2001)[28] and gave Perkin Gilman's work a visual narrative, exploring the role of the women within the institution and Kristeva's theory of intertextuality. Writing from a psychoanalytical, literary and philosophical perspective, Miglena Nikolchina, in *Matricide in Language* (2004), focused on Kristeva's intertextuality when exploring the connections between Virgina Woolf and Kristeva.[29]

The association between poetic images and text has become more prevalent in my artistic practice. Through the visual language that develops from a background of ambiguity within the films, this non-verbal communication translates through the performative use of time-based media. The films and photographic installations endeavour to create unique lyrical dialects of an evolving painterly and sculptural video language through the incorporation of found objects, which are imbued with multiple meanings. The films have tended to have a fixed focus tableau style framing with minimal lighting and effects – preferably daylight. This process also enables a quiet reflective time to think with a multitude of film footage taken (approximately 2-3 hours per session) proceeded by the editing of the films down to approximately 3-10 minutes. From a meditative synthesis of time, slot length, and layering, the interplay of edits and cuts are the grammatical structuring of the work. The visual architecture upon which the images hang is like the phrasing, meter, rhyming of words in a poem. The titles of the films allow stories to interplay and give the viewer access to the artworks, relating to Kristeva's

---

which not only contested the patriarchal order but also the male as being dual voiced in the sense of retaining the wife's role as subordinate, and also renouncing her ability to recognise her own needs, thereby retaining ultimate control. The dialogue confronted language and in doing so produced its own discourse through its intertextuality. Julia Kristeva, *Desire in Language: A Semiotic Approach to Literature and Art* (New York: Columbia University Press, 1980).

[28] Patricia MacKinnon-Day, *Enclosure, Naked Spaces* (installation in Oriel Davies Gallery, Newtown, Powys, Wales, 2001)

[29] Miglena Nikolchina focused on the intertextual reading of Virginia Woolf and Julia Kristeva in order to address the mystery of the persistent suppression of women's contributions to culture. Nikolchina invited the reader to rethink "the controversial place of the maternal in relation to the symbolic, and [showed] how the maternal emerges within writing as a kind of stammering in search of self-expression." Miglena Nikolchina, *Matricide in Language: Writing Theory in Kristeva and Woolf* (New York: Other Press, 2004), 152.

notion of intertextuality and transportation[30] within the format of a short experimental film.

Fig. 1-8: Heald, K. (2013) *Diapason* [film stills], Ellesmere, UK.

Through the "dream films" I am developing an intuitive way to think through the lens, not just filming randomly but exploring colour, line and form to develop an intensity of digging deeper – unravelling a sense of time, reverie or unknown reality. As an experimental filmmaker I endeavour to portray this journey, attempting to reveal and allude to the invisibility of time. The artworks cited emphasize not only the films' poetics but also reveal the scientific techniques of the camera. Achieved by the absence of linear narratives and sound tracks – by sourcing sound

[30] Kristeva's earliest characterizations of intertextuality (for example, "in the space of a text several utterances, taken from other texts, intersect and neutralize each other") associate it with the process of literary allusion. In *Revolution in Poetic Language*, however, a much stronger accent falls on the dynamics of transposition (that is, the unstable process whereby differently realized texts collide in, say, a novel) and Kristeva goes out of her way to disassociate intertextuality from the "study of sources." Because intertextuality is identified in *Revolution in Poetic Language* as a third primary process (primary processes were Freud's list of psychic activities that effect the form and content of dreams), it is equally clear that transposition is designed to give the analyst access to the permutation of subject positions undergone in the production of a work. "Text/Textuality – Julia Kristeva: Textual Productivity And Intertextuality," <http://science.jrank.org/pages /11415/Text-Textuality-Julia-Kristeva-Textual-Productivity-Intertextuality.html.>.

site-specifically and varying the speed by slow motion – such as in the "dream film" *Diapason* (2013).[31] The film is, therefore, a tool that can be applied to time as the telescope and microscope are applied to space (Fig.1-8).

In the sciences, film has proved revolutionary as a way of amalgamating a variety of temporal phenomena and to record objective reality. Slowing down film also brings about its poetics, evoking dreams and entering into the submerged world of the unconscious. By engaging with the Kristevian *chora* the notion of the spoken word, being dissociable yet concentric, has been explored. The artworks became intentionally framed to display and narrate the unconscious recordings and measurements of time, the gap between waking and sleep.

Dreams & Sleep

There is a growing exchange between art and science practitioners and theorists and I have contributed to the dialogues created by previous artists and academics focusing on sleep, time and the politics of location. I have collaborated with artists, sociologists and scientists in Europe and Asia to synthesise women's sleep with time and the *chora* from an arts and science perspective. This has involved employing different approaches, which have varied from empirical and quantifiable scientific testing and analytical processes to qualitative perception, intuition and myth. I have facilitated access to the imaginative dream world, through the creation of "staged" sleep performances in diverse sites and over several phases, which concentrated on female corporeality.

Collaborating with Dr Myriam Kerkoffs, Director of Sleep Research in Belgium on the piece *Drift* (2004-2005)[32] I attempted to analyse time and the unconscious through dialogues with the science specialists by using video equipment common to both disciplines, yet re-initialised by the collaborative arts practice-based investigations and interpretations. My methodology differed in the sense that my research interests were aimed at capturing both an interior and exterior experience of sleep and time, encapsulating not only the dream space but also the social context and the relationship of the body to the site. I explored the notion of transcendence through the "dream films," the social context and psychology of sleep, and considered the very intimate space of sleep and how this could be sensitively addressed through my gaze as a filmmaker (Fig. 2-1).

---

[31] *Diapason,* directed by Karen Heald (UK: Karen Heald, 2013), Video: HD.

[32] *Drift,* directed by Karen Heald (Belgium: Karen Heald, 2004-2005), Video: SD.

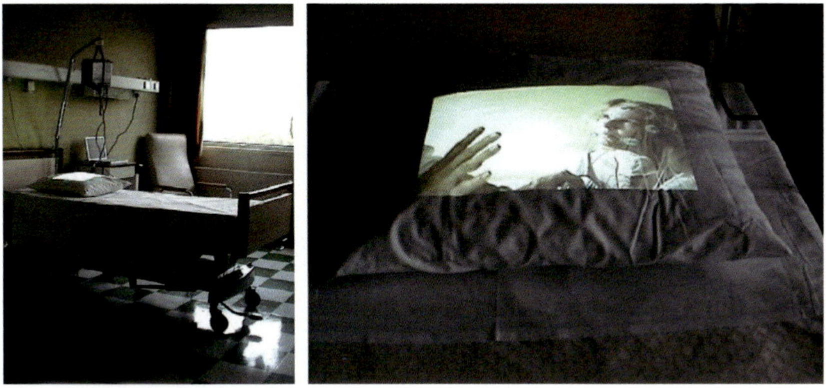

Fig 2-1: Heald, K. (2004-2005) *Drift III* [video installation], Charleroi University Hospital, Belgium.

The ensuing films and images from the exhibitions could be experienced as being extremely personal and yet astonishingly familiar, symptomatic of the shared dream world found in Jung's collective unconscious. Time, as experienced in the *chora* precedes Kristeva's *thetic*[33] phase where the subject is on the threshold of emergence into symbolic language. As such, through the visual and emotional language of "dream films," the "unconscious thetic" latently reveals itself to an enquiring viewer. On completion of *Drift* I was conversant with contemporary artists' and scientists' working in a similar field, such as Catherine Yass and Dr Mary Morrell. I wanted to further investigate the notion of transcendence and the feasibility of encapsulating interior and exterior experiences through the "dream films." Advanced knowledge of the context and psychology of sleep came upon collaborating with sociologist Dr Jenny Hislop and glass artist Chris Bird-Jones on a pilot study resulting in jointly authored papers and exhibitions. Sleep and the female gaze were important issues within the experimental "dream films" in the investigation. Due to often being on both sides of the camera as director and performer I also became the observer and the observed. The interaction with concepts from film theorists and makers, such as Laura Mulvey with her notion of the

[33] Kristeva uses the concept of *chora* to describe the continuous rhythmic and feminine space shared by mother and child prior to the emergence of what she calls the *thetic* phase that prepares the sign for emergence into the symbolic order. Julia Kristeva, *Revolution in Poetic Language* (New York: Columbia University Press, 1974/1984), 25.

"Pensive Spectator" (2006)[34] and Zeinabu Irene Davis' who as, bell hooks, describes is creating an "oppositional gaze" (1992)[35] has enabled me to contextualise the ethics involved in the transition from the camera to the editing suite and from subject to object. Cecily Brennan's film *Melancholia* (2005)[36] was instrumental in the multi-layered relationship

---

[34] Through the "dream films" I began to expand upon the notion of the female gaze within the still and moving image as commented upon by Mulvey in *Death 24x a Second: Stillness and the Moving Image* (2006). Extending upon her arguments from her landmark text *Visual Pleasure and Narrative Cinema* (1975), Mulvey discusses the different types of spectators. Drawing from the work of Raymond Bellour, Mulvey theorises on the 'Possessive Spectator' who is a possessive, "fetishistic" spectator of the new cinema of delay (video). This spectator draws pleasure from the ability to stop, control, repeat, and somehow possess stilled film images, particularly of movie stars. She then provides the idea of the gender neutral 'Pensive Spectator' for whom the cinema of delay prompts considerations about cinema itself and for whom fascination displaces fetishistic scopophilia as the main form of pleasure. Laura Mulvey, *Death 24x a Second: Stillness and the Moving Image* (London: Reaktion Books, 2006), 190-196.

[35] Writing from a black feminist perspective American author, bell hooks, put forth the notion of the "oppositional gaze," encouraging black women not to accept stereotypical representations in film, but rather actively critique them. bell hooks, *Black Looks: Race and Representation* (Boston: South End Press, 1992), 115. Zeinabu Irene Davis, one of a growing number of independent black women filmmakers, has actively constructed this "oppositional gaze" that looks through and at African-American female experiences via a black lens. Davis' film *Co-Motion: Tales of Breastfeeding Woman* seeks to reclaim black female subjectivity through poetic (re)constructions of time, the body, and an exploration of spatial configurations. *Co-Motion: Tales of Breastfeeding Woman*, directed by Zeinabu Irene Davis (United States, UCLA Film and Television Archive, 2010), Film: 16mm.

[36] Visiting Oriel Mostyn's (2011) exhibition *Love Film*, I experienced two films directed by Cecily Brennan, *Melancholia* (Ireland: Cecily Brennan, 2005) and *Unstrung* (Ireland: Cecily Brennan, 2007) which utilized the female body. Observing these films I immediately comprehended and connected to Brennan's more obvious symbols (trestles, black liquid, etc.) through which she referenced Kristeva's *Black Sun: Depression and Melancholia* (New York: Columbia University Press, 1989), and the 'abject' from Kristeva's, *Desire in Language: A Semiotic Approach to Literature and Art.* (New York: Columbia University Press, 1980). Caoimhín Mac Giolla Léith describes this connection to the 'lethal flood' or 'oceanic death' articulated by Kristeva in her account of depression and melancholia as a "traumatic state [which] is, however, given palpable form in the physically inexplicable ooze symbolising black bile, an excess of which the ancients believed to be the cause of melancholia (from Ancient Greek melas, 'black', + kholé, 'bile)." Mac Giolla Léith also discusses Brennan's undeniable

not only with painters such as Hans Holbein and Albert Moore but also Kristeva's publication *Black Sun* (1989),[37] melancholia and the abject. As a female filmmaker and researcher, I attempted to distort the gaze and portray sensitive, non-vouyeristic depictions of women at sleep – a time when they are most vulnerable. Doing this required mutual trust and respect to enable a uniquely female visual language. In transcending verbal communication this paradoxically often involved direct observational scientific methodologies and becoming the researcher and the researched in complex collaborations.

Sleep also encounters the realm of the theatrical performance with the bed as the stage forming a dramaturgical analysis as discussed by Erving Goffman in *The Presentation of Self in Everyday Life* (1956).[38] This related to the "transitory strata" and the transitional spaces between the conscious and the unconscious and it was at this point that I first introduced the diary format as a creative tool into the art practice. This was

---

homage to Hans Holbein's *The Body of Christ in the Tomb* (1521) and states that "unlike Holbein's physically tortured dead Christ, stretched and supine in his ornately carved cask, the body of Brennan's living woman, is cramped and slightly contorted, wracked by a less palpable torment." Caoimhin Mac Giolla Léith, "Damage and Survival," in *Cecily Brennan: Selected Works 1999-2007,* ed. Cecily Brennan (Skibbereen: West Cork Arts Centre, 2007). This relates to Bram Dijkstra's article *Idols of Perversity* (1986) which critiques Albert Moore's painting *Apples* (1875). By depicting two sleeping women and stating that their bodies while in "the process of trying to find a comfortable position … [they were] … trapped in peculiar twists, which emphasized the extent of their languorous passivity ..." Bram Dijkstra, *Idols of Perversity: Fantasies of Feminine Evil in Fin-de-Siècle Culture* (Oxford: Oxford University Press, 1986). However in Brennan's *Melancholia* the woman has her eyes open, yet avoids the viewer's gaze as she frequently glances downwards towards the floor.

[37] Julia Kristeva, *Black Sun: Depression and Melancholia,* trans. Leon. S. Roudiez (New York: Columbia University Press, 1989).

[38] The notion of sleep as a theatrical performance, with the bed being the stage, draws on Canadian sociologist and writer Erving Goffman's study of symbolic interaction in the form of dramaturgical analysis. Goffman framed the symbolic interaction as the theatrical performance that applies to face-to-face interactions, believing that when individuals come into contact with other people, they change their appearance or manner in an attempt to control or guide impressions others may perceive of them. Goffman claimed that privacy lies beneath all social interaction and that the dramaturgical approach functions to explain this mystery. I attempted to investigate this private place by exploring the transitional space between the conscious and unconscious, of "transitory strata," and the temporal boundaries of dormancy in the personal space of the bedroom. Erving Goffman, *The Presentation of Self in Everyday Life* (New York: Doubleday, 1956).

an attempt to record the opportunities and challenges that occurred from the interaction between the sociologist and the artist, the reversal of roles and the spaces that lay "in-between." This motility of space was played out theatrically in the use of the materials chosen with Bird-Jones in *The Pillow Series*. The association was drawn between various fairy tale characters from films and literary fields such as *Sleeping Beauty* (2008) to current artist performances, for example, Cornelia Parker and Tilda Swinton's *The Maybe* at the Museum of Modern Art in New York (2013).[39] However, that is where the resemblance ended, for the women displayed in the glass pillows were not trapped. By filming women and separating the dreamer from the viewer, *The Pillow Series I & II* (2006-2007)[40] investigated what it might mean as women to be "staging" sleep performances, being simultaneously in other worlds and timelines. This evolved "transitory strata" by visualising a re-presentation of sleep, the unconscious and experiential time (Fig.2-2).

Fig. 2-2: Bird-Jones, C. & Heald, K. (2006-2007) *The Pillow Series I & II* [video installation & detail], Kunstraum Gallery, Lüneburg, Germany.

[39] The artist, Cornelia Parker collaborated with actress Tilda Swinton in a performance/installation called *The Maybe* which portrayed Swinton lying fully clothed and apparently asleep, inside a vitrine. Swinton was put on display, but appeared trapped at the same time as the work questioned the role of celebrities in our society. Parker first exhibited *The Maybe* in the Serpentine Gallery, London in 1995 but more recently it was repositioned in the Museum of Modern Art (MOMA) in New York in 2013. Cornelia Parker & Tilda Swinton, *The Maybe* (performance/installation in Museum of Modern Art, New York, USA, 2013).

[40] The idea of making glass pillows to house filmed sleep performances become known as *The Pillow Series*. As a succession the artworks created for *The Pillow Series I & II* were exhibited in various formats at the Kunstraum Gallery in Germany. Chris Bird-Jones & Karen Heald *The Pillow Series I & II* (video installations in Kunstraum Gallery, Lüneburg, Germany, 2006-2007).

The research project, *The Dream Space*,[41] was a natural extension of *The Pillow Series*, and enabled the investigation to evolve through "staging" sleep performances in Asia with international collaborators. In response to Kathleen Kirby (1993) and Adrienne Rich's notions of "locational politics" I questioned what sleep and the body meant cross-culturally and in different places through the art of "dream films." The significance of the relationship between the sleeper and the site became paramount. Travelling from Beijing to Ulaanbaatar on the Trans-Mongolian Railway, I visualised the train as a space between "here, there and nowhere." Recording a "staged" sleep performance in a carriage with Beth Turnbull, a newly acquainted passenger, involved immense trust, as feelings of vulnerability and intimacy are present when people are sleeping. In talking with Beth, our initial conversations included dialogues around RoseLee Goldberg's publication *Performance Art: From Futurism to the Present* (1988)[42] that discussed the place of performance in art history, which started in tribal rituals, medieval passion plays and Renaissance spectacles. For the sleep performance *Beth* (2007)[43] similar to other women that I have filmed, Beth was clothed in her own garments and did not expose her body. The ensuing "dream films" were an attempt to offer observational non-voyeuristic experiences for the viewer/reader (Fig 2-3). By filming one-to-one with my collaborators, the artworks were created in personal rooms or spaces where the person felt relaxed and was able to forget that the camera existed.

Continuing *The Dream Space* in Japan, I explored what it was like to experience a state of "being in-between" whereby I slept in contrasting spaces, such as capsule hotels, which are places of mass consumerism and speed, and Buddhist temples which are spaces of reflection and ritual. This enabled the study to enter further into the discourse surrounding sleep performances, such as those previously cited alongside Sophie Calle's work which adopts a narrative approach with the bed as a central theme. Dreaming allows entrance into simultaneous time zones and my interest in

---

[41] Following *The Pillow Series* in Germany, 2007, I received an award to finance a self-initiated research trip to Mongolia, China and Japan. Titled *The Dream Space*, the research project involved "staging" sleep performances in Asia with international collaborators. This fortuitous trip offered me the opportunity to continue my explorations into intimate spaces of sleep, sleeping in different locations and encountering other cultures.

[42] RoseLee Goldberg, *Performance Art: From Futurism to the Present* (London: Thames & Hudson, 1988).

[43] *Beth*, directed by Karen Heald & Beth Turnbull (China/Mongolia: Karen Heald, 2007), Video: SD.

Fig. 2-3: Heald, K. & Turnbull, B. (2007) *Beth* ["staged" sleep performance], Trans-Mongolian Railway, Beijing to Ulaanbaatar.

the subject culminated in various films circulating at the time such as *Pan's Labyrinth* (2006),[44] *Babel* (2006)[45] and "*Wandafuru Raifu*" (*After Life*)[46] by Hirokazu Kore-Eda (1998/2007). I related them to Akira Kurosawa's film *Dreams* (1990)[47] and Bill Viola's film *Chott el-Djerid* (*A*

[44] *Pan's Labyrinth*, directed by Guillermo del Toro (Spain: Optimum Home Entertainment, 2006), DVD.

[45] *Babel*, directed by Alejandro González Iñárritu (United States/Mexico/France: Paramount Pictures Home Entertainment, 2007), DVD.

[46] *Wandafuru Raifu (After Life)*, directed by Hirokazu Kore-Eda (Japan/UK: Soda Pictures Ltd. Home Entertainment, 2007), DVD.

[47] *Dreams,* directed by Akira Kurosawa (United States: Warner Bros. & Akira Kurosawa USA, 1990), Film: 35mm.

Portrait in Light and Heat) (1979).[48] These multi-narrative film dramas impacted my sense of time, and my relationship to travelling, whether physically or via the imagination through sleeping and dreaming.

The relationship between site and performance in Laurie Anderson's *Institutional Dream Series* (1972-1973)[49] and Tehching Hsieh's *One Year Performance* (1981-1982)[50] was created by the artists sleeping in various locations around New York. These "experiments" parodied empirical research methods as conducted by cognitive psychologists of that time, in an effort to interpret what transpires when we sleep. The artists merged these "invisible" performances with the environment to test the body's stamina and restrictions whilst obscuring its theatricality being outside of the art context. In contrast to Anderson's and Hsieh's performances, the art context in *Drifting on a Pink Cloud*[51] was made evident by the use of the small video equipment where the live and the recorded elements enabled random audience participation and framing of sleep and reverie within the work. The aspiration for the viewer was to create a dichotomy, producing juxtaposing emotions, whereby the sleeping vulnerable woman positioned in the beautiful landscape created a tension in the work.

John Cage (1970)[52] and Neil Mulholland (2009)[53] debate the use of sound within artists' performances and installations as an ambient source. Sounds, or the lack of them, play a significant role within the "dream films" which are frequently silent or incorporate ambient noises such as sounds found within the environment. Occasionally the "dream films" amalgamate vocals and noises created in collaboration with singers and musicians. In *The Last Stop* (2007)[54] the notion of the lullaby was used to mirror the complexity of the *chora* (Fig 2-4).

---

[48] *Chott el-Djerid (A Portrait in Light and Heat),* directed by Bill Viola (United States: Electronic Arts Intermix Home Entertainment, 1979), VHS.

[49] Laurie Anderson, "Institutional Dream Series," (1972-1973), <http://www.ntticc.or.jp/Archive/2005/THERECORDOFTHETIME/Works/index.html>.

[50] Tehching Hsieh, *One Year Performance 1981–1982*, (Sean Kelly Gallery/Museum of Modern Art, January 21-May 18, 2009), <http://www.skny.com /artists/tehching-hsieh/images/>.

[51] *Drifting on a Pink Cloud,* directed by Karen Heald & Rachael Kearney (Japan: Karen Heald, 2007), Video: SD.

[52] John Cage, *Song Books Volume 1: Solos for Voice 3-58* (New York: Edition Peters, 1970).

[53] Neil Mulholland, "Dans l'air du Soir - Notes on Ambient Art," (Public Art Scotland, March, 2009), <http://www.publicartscotland.com/features/9-Dans-l-air-du-Soir-Notes-on-Ambient-Art>.

[54] *The Last Stop,* directed by Karen Heald & Rachael Kearney (Japan: Karen Heald, 2007), Video: SD.

Fig 2-4: Heald, K. & Kearney, R. (2007) *The Last Stop* [video performance], Kyoto, Japan.

These diverse methods create sounds which, along with our dreams, pose a challenge to our rational model of the world as space and time are suspended. Dreams are often silent and this suspension occurs when engrossed in the films with ambient sounds often creating an interruption to the flow of time and space. This interruption within dreams not only occurs within the audience's feelings of ambient sounds but can also be experienced within the use of colour. This feature enables unexpected visual experiences in the moving image as portrayed in *The Red Tories in Conversation with the Gold Shoes* (2007),[55] which is similar to the sudden intrusion of colour in Shirin Neshat's film *Zanan-e bedun-e mardan (Women Without Men,* 2009) (Fig 2-5).[56]

Fig 2-5: Heald K. & Kearney, R. (2007) *The Red Tories in Conversation with the Gold Shoes*, [filmed performance], Kyoto, Japan.

[55] *The Red Tories in Conversation with the Gold Shoes,* directed by Karen Heald & Rachael Kearney (Japan: Karen Heald, 2007), Video: SD.

[56] *Zanan-e bedun-e mardan (Women Without Men)*, directed by Shirin Neshat and Shoja Azari (UK: Artificial Eye, 2009), Film: 35mm.

Continuing from *The Dream Space*, I had an opportunity to participate in the *Dream Director* (2009)[57] at Compton Verney art gallery and question how a location changes the context for the audience. Jerram attempted to offer a new tool for sleep, science and clinical applications and raised questions about the rules of interaction and the boundaries of science and art. Using a range of diverse methods before and after *The Dream Director* I have tested the sleep performances which are an attempt to capture "women's time," the female body and the unconscious through audience reactions to the "dream films." For example in 2013, Bird-Jones and I stayed overnight in a Bed & Breakfast in Wales and the resulting installation involved *The Pillow Series IV* (2013)[58] being exhibited in Oriel Davies Gallery but this time existing inside two glass pillows and presented as a diptych (Fig. 2-6).

Fig. 2-6: Bird-Jones, C. & Heald, K. (2013) *The Pillow Series IV* [video installation & detail], Oriel Davies Gallery, Newtown, Wales.

My intention is not to resolve or affirm any particular position or didactic as I operate between art, science and academia but to embed creative and empirical processes side by side to enable an understanding of a metaphysical sense of time. In doing so, I have discussed sleep as metaphor, as experienced dreams and as conversations around the vulnerability of the female body whilst sleeping. The artwork itself has

[57] Luke Jerram, *The Dream Director* (installation in Compton Verney art gallery, Warwick, Warwickshire, 2009).

[58] Chris Bird-Jones & Karen Heald, *The Pillow Series IV* (video installation in Oriel Davies Gallery, Newtown, Wales, 2013).

explored narratives from complex timelines to investigate preverbal sounds and the sounds of the unconscious to explore cultural language and the resonances of sleep and time. Throughout the research I have created "dream films" to produce ambient environments, hoping to retain the audience at the threshold of the "unconscious thetic."

References

Author Unknown, 2007-2008, *Wellcome Collection. Sleeping and Dreaming,* London: Wellcome Trust, exhibition.

Anderson, L., 1972-1973, *Institutional Dream Series*, <http://www.ntticc.or.jp/Archive/2005/THERECORDOFTHETIME/Works/index.html>, online image.

Aronofsky, D., 2011, *Black Swan*, United States: Fox Searchlight Pictures. Home Entertainment, DVD.

Baring, A., and Cashford, J., 1993, *The Myth of the Goddess*, London: Penguin Books.

Brennan, C., 2005, *Melancholia*, Ireland: Cecily Brennan. DVD.

—. 2007, *Unstrung*, Ireland: Cecily Brennan, DVD.

Butler, A., 2007, "Motor-driven metaphysics: movement, time and action in the films of Maya Deren," *Screen* 48, 1-23.

Cage, J., 1970, *Song Books Volume 1: Solos for Voice 3-58*, New York: Edition Peters.

Campion, J., 2007, *The Piano*, Australia: Cibey 2000. Home Entertainment, DVD.

Carter, A., 2008, *The Fairy Tales of Charles Perrault*, London: Penguin.

Daldry, S., 2003, *The Hours,* United States: Paramount Pictures and Miramax Films. Home Entertainment, DVD.

Davis, Z. I., 2010, *Co-Motion: Tales of Breastfeeding Woman*, United States, UCLA Film and Television Archive, Film: 16mm.

Deren, M., 1946, *Ritual in Transfigured Time*, United States: Howard Gotlieb Archival Research Centre, Boston University, Film:16mm.

Deren, M., and Hamid, A., 1943, *Meshes of the Afternoon,* United States: Howard Gotlieb Archival Research Centre, Boston University, Film: 16mm.

Dijkstra, B., 1986, *Idols of Perversity: Fantasies of Feminine Evil in Fin-de-Siècle Culture*, Oxford: Oxford University Press.

Foster, G. A., 1997, *Women Filmmakers of the African & Asian Diaspora: Decolonizing the Gaze, Locating Subjectivity*, Illinois: Southern Illinois University Press.

Gregory, R. L., 1987, *The Oxford Companion To The Mind*, USA: Oxford University Press.

Goffman, E., 1956, *The Presentation of Self in Everyday Life*, New York: Doubleday.

Goldberg, R., 1988, *Performance Art: From Futurism to the Present*, London: Thames & Hudson.

Heald, K., *Diapason,* 2013, UK: Karen Heald, Video: HD.

—. 2014, *"Dream Films" and Research as Collaborative Practice through Contemporary Art and Science Methodologies*, Leeds: Leeds Metropolitan University.

—. *Drift,* 2004-2005, Belgium: Karen Heald, Video: SD.

—. *pra II*, 2005, Czech Republic/UK: Karen Heald, Video: SD.

Heald, K., and Hislop, J., 2007, "Intimate Space: Sleep, Art and Sociology," paper presented at the annual meeting for the European Sociological Association (ESA) Research Network, Luneburg, Germany, March 28-April 1, 2007.

Heald, K., and Kearney, R., 2007, *Drifting on a Pink Cloud,* Japan: Karen Heald, Video: SD.

—. 2007, *The Last Stop,* Japan: Karen Heald, Video: SD.

—. 2007, *The Red Tories in Conversation with the Gold Shoes,* Japan: Karen Heald, Video: SD.

Heald, K., and Liggett, S., 2013, *Paper Interior*, Wales: Karen Heald, Video: HD.

—. 2009, *The Artists' Creation,* Wales: Karen Heald, Video: HD.

—. 2013, *White*, Wales: Karen Heald, Video: HD.

Heald, K., and Sabin, C., 2010, *Not still at home: equilibrium,* Wales: Karen Heald, Video: HD.

—. 2010, *Not still at home: poise*, Wales: Karen Heald, Video: HD.

Heald, K., and Turnbull, B., 2007, *Beth*,. China/Mongolia: Karen Heald, Video: SD.

hooks, b., 1992, *Black Looks: Race and Representation,* Boston: South End Press.

Humm, M., 1995, *The Dictionary of Feminist Theory*, Hemel Hempstead: Prentice Hall/ Harvester Wheatsheaf.

Iñárritu, A. G., 2006, *Babel*, United States/Mexico/France: Paramount Pictures. Home Entertainment, DVD.

Jerram, L., 2009, *The Dream Director*, Warwick, Warwickshire: Compton Verney art gallery, installation.

Jung, C. G., 1954, *The Archetypes and the Collective Unconscious*, Hull, R. F. C., trans., London: Routledge.

Kirby, K. M., 1993, "Thinking through the Boundary: The Politics of Location, Subjects, and Space," *Boundary*, no. 2, 173-189.

Kore-Eda, H., 2007, *Wandafuru Raifu (After Life)*, Japan/UK: Soda Pictures Ltd. Home Entertainment, DVD.

Kristeva, J., 1989, *Black Sun: Depression and Melancholia*, Roudiez, L. S., trans., New York: Columbia University Press.

—. 1980, *Desire in Language: A Semiotic Approach to Literature and Art,* Gora, T., and Jardine, A., trans., New York: Columbia University Press.

—. 2001, *Melanie Klein*, Guberman, R., trans., New York: Columbia University Press.

—. 1974/1984, *Revolution in Poetic Language*, Waller, M., trans., New York: Columbia University Press.

—. 1979, "Women's Time", in *New Maladies of the Soul,* Guberman, R., trans., Kristeva, J., ed., New York: Columbia University Press, 201-224.

Kurosawa, A., 1990, *Dreams,* United States: Warner Bros. & Akira Kurosawa USA, Film: 35mm.

Lechte, J., 1994, *Fifty Key Contemporary Thinkers: from Structuralism to Postmodernity*, London: Routledge.

Mac Giolla Léith, C., 2007, "Damage and Survival," in Brennan, C., ed., *Cecily Brennan: Selected Works 1999- 2007*, Skibbereen: West Cork Arts Centre.

MacKinnon-Day, P., 2001, *Enclosure, Naked Spaces*, Newtown, Powys: Oriel Davies Gallery, installation.

Metz, Ch., 1982, *The Imaginary Signifier: Psychoanalysis and the Cinema*, Indiana: Indiana University Press.

Minh-ha T. T. and Bourdier, J.-P., 2001, *The Fourth Dimension*, Japan: Women Make Movies. Home Entertainment, DVD.

Neil Mulholland, 2009, "Dans l'air du Soir – Notes on Ambient Art," Public Art Scotland, <http://www.publicartscotland.com/features/9-Dans-l-air-du-Soir-Notes-on-Ambient-Art>.

Mulvey, L., 2006, *Death 24x a Second: Stillness and the Moving Image*, London: Reaktion Books.

—. 1975, "Visual Pleasure and Narrative Cinema," *Screen* 16, 6-18.

Neshat Sh., 2001, "Men Without Women/Zanan-e bedun-e mardan," <http://www.imdb.com/title/tt1498887/?ref_=nm_ov_bio_lk1>.

Neshat, Sh., and Azari, Sh., 2009, *Zanan-e bedun-e mardan (Women*

Without Men), UK: Artificial Eye, Film: 35mm.

Nikolchina, M., 2004, *Matricide in Language: Writing Theory in Kristeva and Woolf*, New York: Other Press.

Toro, G. del, 2006, *Pan's Labyrinth*, Spain: Optimum. Home Entertainment, DVD.

Parker, C., and Swinton, T., 2013, *The Maybe*, New York: Museum of Modern Art, performance/installation.

Perkins Gilman, C., 1892, "The Yellow Wallpaper. A Story," *The New England Magazine*, no. 11, 5, 647-657.

Plato, 2008, *Timaeus and Critias*, Lee, D., and Kjeller Johansen, T., trans., London: Penguin Classics.

Rich, A., 1984, *Notes Towards a Politics of Location in Blood, Bread and Poetry: Selected Prose 1979-85*, New York: W. W. Norton & Company.

"Text/Textuality – Julia Kristeva: Textual Productivity And Intertextuality," Science.jrank.org, <http://science.jrank.org/pages/11415/Text-Textuality-Julia-Kristeva-Textual-Productivity-Intertextuality.html>.

Tehching, H., 2009, *One Year Performance 1981–1982*, Sean Kelly Gallery/Museum of Modern Art, <http://www.skny.com/artists/tehching-hsieh/images/>.

Viola, B., 1979, *Chott el-Djerid (A Portrait in Light and Heat),* United States: Electronic Arts Intermix. Home Entertainment, VHS.

Woolf, V., 1926, *Mrs Dalloway*, London: Penguin Classics.

NOTES ON CONTRIBUTORS

Katarzyna Ancuta is a lecturer at the Graduate School of English at Assumption University in Bangkok, Thailand. Her research interests tend to focus on contemporary cultural manifestations of the Gothic, particularly in such areas as popular fiction, film and video, multimedia and performance art, theatre, music and dance, comics and graphic novels, fashion and alternative lifestyles. Most of her publications are concerned with the interdisciplinary contexts of contemporary Gothic and Horror, and recently with (South)-East Asian (particularly Thai) cinema and supernatural anthropology. She is currently working on a book on Asian Gothic, and a multimedia project on Bangkok Gothic. Her recent involvement with the Gothic academic community includes cooperation on *The Encyclopedia of the Gothic* (ed. David Punter, William Hughes and Andrew Smith, Wiley-Blackwell, 2013), The Blackwell *A New Companion to the Gothic* (ed. David Punter, Wiley-Blackwell, 2012), *The Cambridge Companion to the Modern Gothic* (ed. Jerrold E. Hogle, upcoming), and a book on Global Gothic (ed. Glennis Byron, 2013).

Tomasz Burzyński received his PhD in 2009 from the University of Silesia, Poland. He works at the Institute of English Cultures and Literatures where he teaches Cultural Studies, Sociology of Culture and Media Studies. His research interests include cultural studies and cultural theory, cultural theories of risk, and media studies.

Alicja Bemben is a Ph.D. student in the Institute of English Cultures and Literatures at the University of Silesia. Her research interests include philosophy of history, historiography and writings of Robert Graves. She is a member of the Robert Graves Society and participated, among others, in [Re]visions of History in Language and Fiction, Robert Graves and Modernism, and TimeLing conferences.

Sonia Front is an Assistant Professor at the University of Silesia, Poland, where she teaches contemporary British literatures in English. She has published on postmodernist British literature and contemporary film. Her main publications are *Transgressing Boundaries in Jeanette Winterson's Fiction* (Peter Lang, 2009) and a co-edited volume *Interiors: Interiority/*

Exteriority in Literary and Cultural Discourse (Cambridge Scholars, 2010). Currently she is working on a monograph dealing with time in twenty-first-century British quantum fiction.

Karen Heald is an artist, filmmaker and an Honorary Research Fellow in the School of Social Sciences at Bangor University, UK. She holds a practice-based PhD and an MA in Contemporary Fine Art from Leeds Metropolitan University, UK. Her research involves a complex collaborative methodology that has been established through the performative use of time-based media while engaging with the psychoanalyst, linguist and philosopher, Julia Kristeva's concept of the semiotic chora. Her working methodology draws on this notion as a preverbal and creative space that relates to rhythms, colours and trace. She has identified a number of new concepts: "transitory strata", the "unconscious thetic" and "in-between-ness". Through the unique use of the video camera, the editing process and analogies and cross-readings between theory and practice, she creates experimental "dream films" which explore non-linear preverbal spaces and visualize a re-interpretation of sleep, the unconscious and experiential time. Her intention for the viewer on experiencing the "dream films" is to offer a "translation" of how we may encounter and be affected by time. And time, in all its varieties, is here intended as a mechanism through which to formulate significant artistic methodologies on the "in-between-ness" of filmmaking and research as collaborative practice.

Sławomir Konkol is a teacher at the University of Silesia and the University of Bielsko-Biała, PhD student at the University of Silesia, interested in contemporary British literature, American literature and culture, psychoanalysis and masculinity studies. Recent publications include "What Else Is Civilisation For? Narration Overcoming Fear and Trauma in Graham Swift" in *Fearful Symmetries: Representations of Anxiety in Cultural, Literary and Political Discourses*, ed. Leszek Drong, Jacek Mydla and "Assuming We Can Give Some Meaning to the Word 'Nature': Graham Swift via Jacques Lacan" in *Nature(s): Environments We Live By in Literary and Cultural Studies*, ed. Tomasz Gnat, Jacek Mydla, Agata Wilczek.

Jacek Mydla is an Associate Professor at the Institute of English Cultures and Literatures (University of Silesia, Katowice and Sosnowiec, Poland). Currently, he is Director of the Institute. He conducts research and lectures in the history of British literature, specifically British drama (Shakespeare through romanticism), Gothic fiction and drama, and theory of narrative. His book-length publications are: the Dramatic Potential of Time in Shakespeare and Spectres of Shakespeare, a study of appropriations of Shakespeare's drama by early English Gothic authors and playwrights. Also, recently he has published a collection of essays on the dynamics of human time in Shakespeare. In his recent articles, in Polish and English, Mydla has been concerned with romantic drama (e.g. Scottish playwright Joanna Baillie), aspects of British empiricism in the eighteenth century, and the supernatural in fiction.

Katarzyna Nowak-McNeice is an Assistant Professor at the Department of English of the University of Wroclaw (Poland), where she teaches American Literature and Culture. She obtained her doctoral degree from the University of Wroclaw in 2005. In 2007 she published a monograph titled *Melancholic Travelers: Autonomy, Hybridity and the Maternal* (Peter Lang); in 2010 she co-edited a collection *Interiors: Interiority/ Exteriority in Literary and Cultural Discourse* (Cambridge Scholars). Her research interests include American literature, gender studies, and opera studies.

Agnieszka Podruczna is a PhD student and a lecturer at the University of Silesia at the Department of Postcolonial Studies and Travel Writing, and she is currently in the process of writing her PhD dissertation on the subject of the body in postcolonial science fiction. Her academic interests include first and foremost postcolonial studies, gender studies and the theory of science fiction, but she is also interested in popular culture, queer studies and the theory of postmodernism.

INDEX